Geometric Gems:
An Appreciation for Geometric Curiosities
Volume I: The Wonders of Triangles

T0320715

Problem Solving in Mathematics and Beyond

Print ISSN: 2591-7234
Online ISSN: 2591-7242

Series Editor: Dr. Alfred S. Posamentier
Distinguished Lecturer
New York City College of Technology - City University of New York

---

There are countless applications that would be considered problem solving in mathematics and beyond. One could even argue that most of mathematics in one way or another involves solving problems. However, this series is intended to be of interest to the general audience with the sole purpose of demonstrating the power and beauty of mathematics through clever problem-solving experiences.

Each of the books will be aimed at the general audience, which implies that the writing level will be such that it will not engulfed in technical language — rather the language will be simple everyday language so that the focus can remain on the content and not be distracted by unnecessarily sophiscated language. Again, the primary purpose of this series is to approach the topic of mathematics problem-solving in a most appealing and attractive way in order to win more of the general public to appreciate his most important subject rather than to fear it. At the same time we expect that professionals in the scientific community will also find these books attractive, as they will provide many entertaining surprises for the unsuspecting reader.

Published

Vol. 32 *Geometric Gems: An Appreciation for Geometric Curiosities*
Volume I: The Wonders of Triangles
by Alfred S Posamentier and Robert Geretschläger

Vol. 31 *Engaging Young Students in Mathematics through Competitions —*
World Perspectives and Practices
Volume III — Keeping Competition Mathematics Engaging in
Pandemic Times
Edited by Robert Geretschläger

Vol. 30 *Sharpening Everyday Mental/Thinking Skills Through Mathematics*
Problem Solving and Beyond
by Alfred S Posamentier and Hans Humenberger

For the complete list of volumes in this series, please visit www.worldscientific.com/series/psmb

Problem Solving in
Mathematics and Beyond

Volume **32**

Geometric Gems:
An Appreciation for Geometric Curiosities
Volume I: The Wonders of Triangles

Alfred S. Posamentier
Robert Geretschläger

 **World Scientific**

NEW JERSEY · LONDON · SINGAPORE · BEIJING · SHANGHAI · HONG KONG · TAIPEI · CHENNAI · TOKYO

Published by

World Scientific Publishing Co. Pte. Ltd.

5 Toh Tuck Link, Singapore 596224

USA office: 27 Warren Street, Suite 401-402, Hackensack, NJ 07601

UK office: 57 Shelton Street, Covent Garden, London WC2H 9HE

Library of Congress Cataloging-in-Publication Data
Names: Posamentier, Alfred S., author. | Geretschläger, Robert, author.
Title: Geometric gems : an appreciation for geometric curiosities /
    Alfred S. Posamentier, Robert Geretschläger.
Description: New Jersey : World Scientific, [2024]- |
    Series: Problem solving in mathematics and beyond, 2591-7234 ; volume 32 |
    Includes index. | Contents: Volume 1. The wonders of triangles --
Identifiers: LCCN 2023043230 | ISBN 9789811279584 (vol. 1 ; hardcover) |
    ISBN 9789811281914 (vol. 1 ; paperback) | ISBN 9789811279591
    (vol. 1 ; ebook for institutions) | ISBN 9789811279607 (vol. 1 ; ebook for individuals)
Subjects: LCSH: Geometry--Popular works. | Geometry, Plane--Popular works. |
    Triangle--Popular works.
Classification: LCC QA445 .P668 2024 | DDC 516--dc23/eng/20231023
LC record available at https://lccn.loc.gov/2023043230

British Library Cataloguing-in-Publication Data
A catalogue record for this book is available from the British Library.

For any available supplementary material, please visit
https://www.worldscientific.com/worldscibooks/10.1142/13507#t=suppl

Desk Editor: Rosie Williamson

Project Managed and Typeset by The Froebe Group/Manila Typesetting Company

Printed in Singapore

Contents

Acknowledgments

It is well known that mathematical problems often lend themselves to alternative solutions. Our goal in this book has been to provide the most elegant and efficient proofs to some of the most challenging geometric problems. Towards this end, several mathematicians have provided us with some wonderful ideas. These include David Hankin, Moritz Hiebler, Hans Humenberger, and Robert Serkey. We thank them for sharing their brilliance with us.

About the Authors

Alfred S. Posamentier

Alfred S. Posamentier is currently Distinguished Lecturer at New York City College of Technology of the City University of New York. Prior to that he was Executive Director for Internationalization and Funded Programs at Long Island University, New York. This was preceded by five years as Dean of the School of Education and Professor of

Mathematics Education at Mercy University, New York. Before that he was for 40 years at The City College of the City University of New York, at which he is now Professor Emeritus of Mathematics Education at and Dean Emeritus of the School of Education. He is the author and co-author of more than 80 mathematics books for teachers, secondary and elementary school students, as well as the general readership. Dr. Posamentier is also a frequent commentator in newspapers and journals on topics related to education and mathematics.

After completing his B.A. degree in mathematics at Hunter College of the City University of New York, he took a position as a teacher of mathematics at Theodore Roosevelt High School (Bronx, New York), where he focused his attention on improving the students' problem-solving skills and at the same time enriching their instruction far beyond what the traditional textbooks offered. During his six-year tenure there, he also developed the school's first mathematics teams (both at the junior and senior level). He is still involved in working with mathematics teachers and supervisors, nationally and internationally, to help them maximize their effectiveness.

Immediately upon joining the faculty of the City College of New York in 1970 (after having received his master's degree there in 1966), he began to develop in-service courses for secondary school mathematics teachers, including such special areas as recreational mathematics and problem solving in mathematics. As Dean of the City College School of Education for ten years, his scope of interest in educational issues covered the full gamut educational issues. During his tenure as dean he took the School from the bottom of the New York State rankings to the top with a perfect NCATE accreditation assessment in 2009. He also raised more than 12 million dollars from the private sector for educational innovative programs. Posamentier repeated this successful transition at Mercy University, where he enabled it to become the only college to have received both NCATE and TEAC accreditation simultaneously.

In 1973, Dr. Posamentier received his Ph.D. from Fordham University (New York) in mathematics education and has since extended his reputation in mathematics education to Europe. He has been visiting professor at several European universities in Austria,

England, Germany, Czech Republic, Turkey and Poland. In 1990, he was Fulbright Professor at the University of Vienna.

In 1989 he was awarded an Honorary Fellow position at the South Bank University (London, England). In recognition of his outstanding teaching, the City College Alumni Association named him Educator of the Year in 1994, and in 2009. New York City had the day, May 1, 1994, named in his honor by the President of the New York City Council. In 1994, he was also awarded the *Das Grosse Ehrenzeichen für Verdienste um die Republik Österreich*, (Grand Medal of Honor from the Republic of Austria), and in 1999, upon approval of Parliament, the President of the Republic of Austria awarded him the title of *University Professor of Austria*. In 2003 he was awarded the title of *Ehrenbürgerschaft* (Honorary Fellow) of the Vienna University of Technology, and in 2004 was awarded the *Österreichisches Ehrenkreuz für Wissenschaft & Kunst 1.Klasse* (Austrian Cross of Honor for Arts and Science, First Class) from the President of the Republic of Austria. In 2005 he was inducted into the Hunter College Alumni Hall of Fame, and in 2006 he was awarded the prestigious *Townsend Harris Medal* by the City College Alumni Association. He was inducted into the New York State Mathematics Educator's Hall of Fame in 2009, and in 2010 he was awarded the coveted *Christian-Peter-Beuth Prize* from the Technische Fachhochschule – Berlin. In 2017, Posamentier was awarded *Summa Cum Laude nemmine discrepante,* by the Fundacion Sebastian, A.C., Mexico City, Mexico.

He has taken on numerous important leadership positions in mathematics education locally. He was a member of the New York State Education Commissioner's Blue Ribbon Panel on the Math-A Regents Exams, and the Commissioner's Mathematics Standards Committee, which redefined the Mathematics Standards for New York State, and he also served on the New York City schools' Chancellor's Math Advisory Panel.

Dr. Posamentier is still a leading commentator on educational issues and continues his long-time passion of seeking ways to make mathematics interesting to both teachers, students and the general public, as can be seen from some of his more recent books.

For more information and a list of his publications see: https://en.wikipedia.org/wiki/Alfred_S._Posamentier

Robert Geretschläger

Robert Geretschläger is a teacher of mathematics and geometry, living in Graz, Austria. He retired from his teaching position at Bundesrealgymnasium Keplerstraße in Graz in 2022, after working there for nearly 40 years. He is currently still active as a geometry lecturer in teacher education at the University of Graz, Austria.

Born in Toronto, Canada in 1957, he moved to Austria in 1972, where he completed his education, certified with a Magister degree as a secondary school mathematics teacher with a specialty in geometry and subsequently earned a doctorate in Mathematics from the University of Graz.

For most of his career, he has been actively involved in the organization of mathematics competitions at all levels of age and complexity. Among other roles in this context, he has been leader of the Austrian team at the International Mathematical Olympiad since 2007 and the leading organizer of the Mathematical Kangaroo Competition in Austria since its beginning in the mid-1990s. He was also responsible for introducing the *International Mathematical Tournament of the Towns* and the *Mediterranean Mathematics Competition* to Austria.

Internationally, since 2008, he has been a member of the Executive of the *World Federation of Mathematics Competitions* (WFNMC), serving as Senior Vice-President from 2018 to 2022 and as President since 2022. He is also a long-standing member of the international Board of the *Association Kangourou sans Frontières* (AKSF), where he is currently serving as treasurer. He has also lectured at many international venues, having been an invited lecturer in Canada, Australia, Czech Republic, Switzerland and Germany, among other countries.

He has also been involved in a number of mathematics education projects over the years. Among other things, he was a member of the Didactics Commission of the *Österreichische Mathematische Gesellschaft* (ÖMG) for many years and leading editor of a series of high school mathematics books. He was actively involved in curriculum development for the Austrian high schools and played an active role in the early phase of introducing the *PISA* study to Austria.

He is the editor and co-author of numerous books on popular mathematics, problem-solving (competition) mathematics and his special research interest, the Geometry of Origami.

More information is available at his website at www.rgeretschlaeger.com

Introduction

Although we are all exposed to geometric principles in early schooling, not all of us are given the opportunity to truly appreciate the full power and beauty of geometry. Considered either from a purely aesthetic standpoint or as a logical subject, geometry can be considered a key to the appreciation of mathematics. In the United States, the exposure that students get by studying geometry for one entire year in secondary school is certainly enlightening. However, with so much time spent developing an understanding for the basic concepts and working with the underlying principles of geometry, there is little time left to learn to truly appreciate some of the wonders that the subject offers. In other countries, the study of elementary geometry has been radically reduced in recent decades, and appreciation of its beauty has suffered accordingly. Even in countries with a strong tradition of teaching geometry in school, the emphasis tends to be more on technical aspects and logic, often to the detriment of the recognition of its visual splendor.

There are many astonishing and surprising results to be found when geometric relationships are explored. Many of these are quite counterintuitive but astounding nevertheless. A simple example, sometimes presented in secondary school geometry courses, is the fact that consecutively joining the midpoints of the sides of any randomly drawn quadrilateral always yields a parallelogram. There are countless such unexpected curiosities that result from rather simple geometric formations. This book will take motivated readers on a journey through many previously unexplored relationships that will leave them in total amazement. Also, to enlighten the readers, Euclidean proofs are provided for all of these astonishing results. These proofs are presented in as reader-friendly a fashion as possible. Of course, many readers may be content just to marvel at the results without concerning themselves with the proofs. This is fine, as understanding the proofs is in no way required to appreciate these lovely properties.

This is the first of a series of three books exhibiting the splendors of geometry and focuses on triangle relationships. Subsequent books with the same basic goals will focus on quadrilaterals and circles. In order to benefit the most from reading this book, the reader should have a reasonable familiarity with the basic concepts taught in secondary school geometry. When there are geometric theorems used in some of the proofs that may be considered beyond the scope of secondary school geometry, we present them in a Toolbox section, which will elucidate them for the general readership. One should not think that these Toolbox items are all beyond the high school curriculum, as a good portion of this section reviews many of the basics from high school geometry courses as well. In short, the Toolbox provides all the necessary equipment for a reader to understand the proofs offered in the second section of this book.

To make your journey through the book as enjoyable and as entertaining as possible, the organization of the individual examples is rather randomly dispersed, although there are several related sequences intentionally organized to show the development of certain concepts. Some of these sequential units deal with collinearity of points, concurrency of lines, concyclic points, and other topics typically not emphasized in the secondary school geometry curriculum.

The examples selected throughout this book are quite uncommon and have either been self-created or else presented in geometry books over the past two centuries. Readers are advised to first read and truly appreciate the amazing relationships that each of these examples presents. As a next step, an interested reader might focus on proving these astonishing curiosities, especially after experiencing a genuine appreciation of their wonders. To enhance the appreciation of these, we have done our best to present all proofs and enhancements in a reader-friendly fashion.

We invite you now to embark on this marvelous journey through many hidden features of triangle-related geometry that expose the power and beauty of geometry as well as the logic that supports it. Enjoy!

Geometric Curiosities: Introducing the Triangle

Perhaps the most basic structure in linear geometry is the triangle. As we begin our journey through the wonders of the triangle, we should start by introducing the various kinds of triangles and some significant points related to these marvelous figures. When a triangle has no sides of equal length, it is referred to as a *scalene triangle*. A triangle that has two equal sides is referred to as an *isosceles triangle* and a triangle with three equal sides is an *equilateral triangle*. Triangles are also sometimes referred to by their angle size, such as a *right triangle*, which has one angle of 90°, or an *acute triangle*, which has all angles less than 90°. A triangle that has one angle greater than 90° is called an *obtuse triangle*. We now have the basic nomenclature so that we can move along expeditiously.

There are particularly prominent points in every triangle. There is, for instance, the point of intersection of the angle bisectors of a triangle, which is also the center of the inscribed circle of the triangle; therefore, it is called the *incenter* of the triangle. Another significant point in a triangle is the point of intersection of its altitudes, which is called the *orthocenter* of the triangle. Then there is the point of intersection of the medians of a triangle, which is known as its *centroid*, or the balancing-point of the triangle. The point of intersection of the perpendicular bisectors of each of the sides of the triangle is the center of the circumscribed circle of the triangle and is called its *circumcenter*.

Curiosity 1. Angles at the Incenter

Triangle *ABC* in Figure 1 has angle bisectors *AX*, *BY*, and *CZ*, meeting at point *I*, which, as noted above, is the center of the inscribed circle and is referred to as the *incenter* of the triangle. A not well-known curiosity that evolves from the angle bisectors is that the angle formed by two angle bisectors is equal to 90° plus one-half the measure of the third angle.

In Figure 1, for example, $\angle BIC = 90° + \frac{1}{2}\angle A$.

Similarly, $\angle CIA = 90° + \frac{1}{2}\angle B$, and $\angle AIB = 90° + \frac{1}{2}\angle C$.

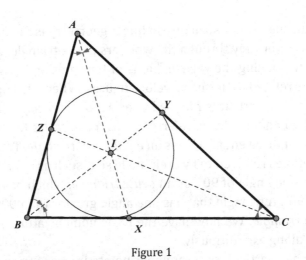

Figure 1

Curiosity 2. An Unexpected Perpendicularity

Let us continue with a rather unexpected curiosity demonstrating some of the wonders that triangles provide. We begin with a randomly chosen triangle *ABC* in Figure 2, whose inscribed circle with center *I* is tangent to the sides of the triangle at points *D*, *E*, and *F*. Line *BI* extended intersects *DE* at point *P*. Quite surprisingly, we find that *AP* is always perpendicular to *BP*.

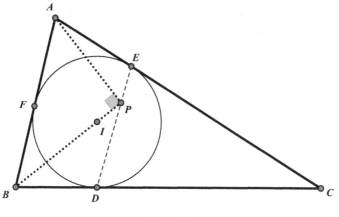

Figure 2

Curiosity 3. A Most Unusual Line Bisection

The lines joining vertices of a triangle with the points of tangency of the inscribed circle can be divided in half in the following way. In Figure 3, circle I is inscribed in triangle ABC at points D, E, and F. The midpoints of sides BC, CA, and AB are points L, M, and N, respectively. When we draw lines IL, IM, and IN, we find, quite unexpectedly, that they bisect each of the lines AD, BE, and CF, at points X, Y, and Z, respectively.

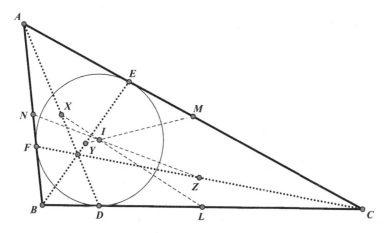

Figure 3

Curiosity 4. Gergonne's Discovery Involving the Inscribed Circle of a Triangle

Continuing with the inscribed circle of a triangle, we can now identify an intriguing point of concurrency. In Figure 4, when we draw the lines from each vertex of triangle *ABC* to the points of tangency, *D*, *E*, and *F*, on the opposite sides, these three lines are concurrent at a point *G*. This point is named the *Gergonne point* of *ABC* after the French mathematician Joseph-Diaz Gergonne (1771–1859). The triangle *DEF*, formed by joining these three points of tangency, is called the *Gergonne triangle*.

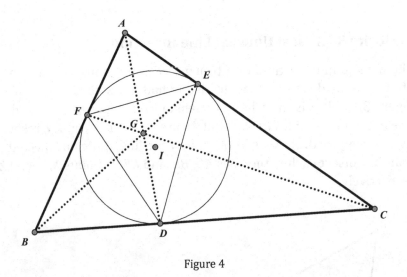

Figure 4

Curiosity 5. Another Unexpected Concurrency for Gergonne's Triangle

There is more of interest that can be found in this configuration. If we draw the perpendicular line segments from the midpoints *L*, *M*, and *N*, of each of the sides of triangle *ABC*, to each of the sides of the Gergonne triangle *DEF*, we find that these line segments, *LX*, *MY*, and *NZ*, are concurrent at point *P*, as we see Figure 5.

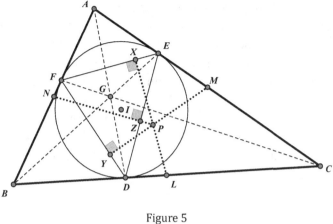

Figure 5

Curiosity 6. A Novelty Concerning the Circumscribed and Inscribed Circles

By considering the inscribed circle and the circumscribed circle of a given triangle, unexpected wonders appear. Many concurrencies and

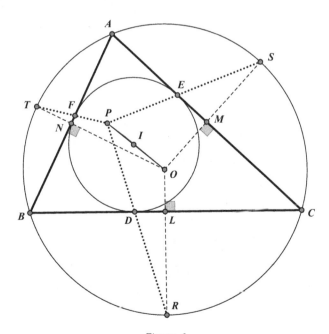

Figure 6

collinearities will surprise us in this configuration. For example, consider Figure 6, where O is the circumcenter of triangle *ABC*. We draw the perpendicular bisectors *OL*, *OM*, and *ON*, of each of the sides of triangle *ABC*, which meet the circumscribed circle at points R, S, and T, respectively, which are the midpoints of the arcs $\overset{\frown}{BC}$, $\overset{\frown}{AC}$, and $\overset{\frown}{AB}$, respectively, determined by the sides of the triangle. We then draw the lines connecting these arc midpoints with the points of tangency of the inscribed circle of triangle *ABC*. We find that these lines *DR*, *ES*, and *FT* are concurrent in a point *P*. Furthermore, another curious result is that the points *P* and *O* are collinear with the incenter *I* and the orthocenter *G*.

Curiosity 7. Concurrency Involving the Inscribed Circle

The inscribed circle can reveal a variety of unexpected astonishments. One such is a concurrency that can be found by taking the diametrically opposite points, R, S, and T, from the points of tangency, D, E, and F, of the inscribed circle of triangle *ABC*, as we see in Figure 7. When we draw the extended lines *AR*, *BS*, and *CT*, we see that these

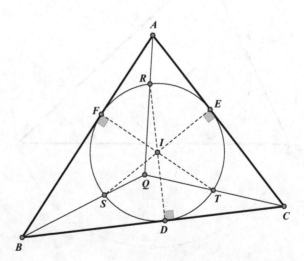

Figure 7

are concurrent at point *Q*, and so, another unexpected concurrency has been discovered.

Curiosity 8. A More General Concurrency Involving the Inscribed Circle

The previous example is actually just a special case of a more general result. In Figure 8, we choose any point *P* in the interior of the inscribed circle of triangle *ABC*, and from the points of tangency, *D*, *E*, and *F*, we draw lines through point *P*, intersecting the inscribed circle at points *R*, *S*, and *T*, respectively. Once again, when we draw the extended lines *AR*, *BS*, and *CT*, we see that these are concurrent at point *Q*. There seems to be an endless number of concurrencies to be discovered in this configuration!

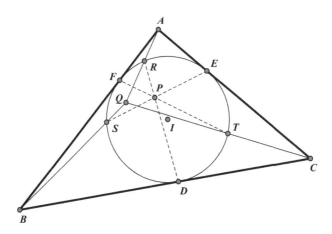

Figure 8

Curiosity 9. An Inscribed Circle of a General Triangle Generates Equal Angles

In Figure 9, circle *I* is inscribed in triangle *ABC*, and *AI* extended meets side *BC* at point *D*. The line segment *IE* is perpendicular to *BC* at point *E*. We then find that ∠*BID* = ∠*EIC*.

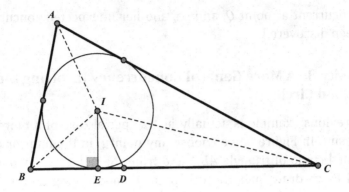

Figure 9

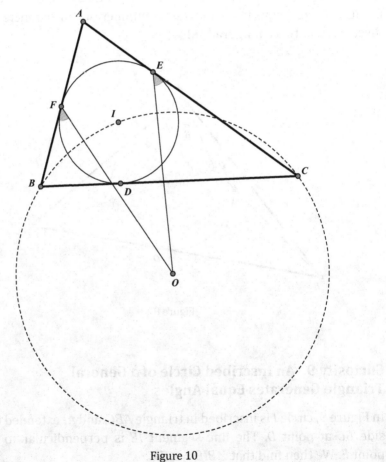

Figure 10

Curiosity 10. Surprising Equal Angles

The circle with center *I* is inscribed in triangle *ABC* and is tangent to the sides of the triangle at points *D*, *E*, and *F*, as shown in Figure 10. Point *O* is the center of the circle containing points *B*, *I*, and *C*. With this simple arrangement, we find quite unexpectedly that $\angle BFO = \angle OEC$.

Curiosity 11. An Inscribed Circle of a Triangle Generates its Circumscribed Circle

We will now consider how perpendiculars related to an inscribed circle of a triangle can generate the circumscribed circle of the same triangle. We begin with triangle *ABC*, as shown in Figure 11, and its inscribed circle with center *I*. We then draw lines *AI*, *BI*, and *CI*, and

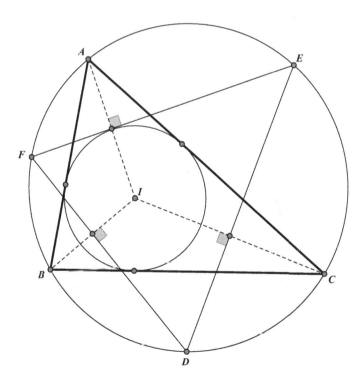

Figure 11

their perpendicular bisectors, so that $EF \perp AI$, $FD \perp BI$, and $DE \perp CI$. The surprising result is that the points D, E, and F all lie on the circumscribed circle of triangle ABC. This is also significant because it is always a special feature to get more than three specific points on the same circle, and in this case, we even have six concyclic points, namely, A, B, C, D, E, and F.

Curiosity 12. The Feet of the Altitudes Partition a Triangle into Three Pairs of Equal-Area Triangles

Points D, E, and F are the feet of the altitudes of triangle ABC, as shown in Figure 12. Point O is the center of the circumscribed circle of triangle ABC. The lines OA, OF, OB, OD, OC, and OE partition the triangle into six smaller triangles, where we find curiously that

$$\text{area}[AOF] = \text{area}[COD], \text{area}[BOF] = \text{area}[COE], \text{area}[BOD]$$
$$= \text{area}[AOE].$$

They may not appear to be equal in area, but, in fact, they truly are.

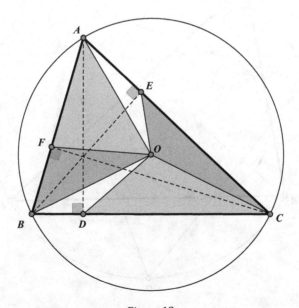

Figure 12

Curiosity 13. The Noteworthy Position of the Orthocenter of a Triangle

The point of intersection of the three altitudes of a triangle, the orthocenter, is inside the triangle if the triangle is an acute triangle, and outside the triangle if it is an obtuse triangle. In Figure 13(a), we show an acute triangle *ABC*, with its three altitudes: *AD*, *BE*, and *CF*, intersecting at point *H*, the *orthocenter* of the triangle. In Figure 13(b), we show an obtuse triangle with its three concurrent altitudes, where the orthocenter, *H*, is outside the triangle.

The orthocenter of a triangle has an unexpected feature in that it partitions each altitude so that the product of the segments of each altitude is equal to the product of the segments of each of the other altitudes. We can see this in Figures 13(a) and 13(b) for triangle *ABC*, where $AH \cdot DH = BH \cdot EH = CH \cdot FH$.

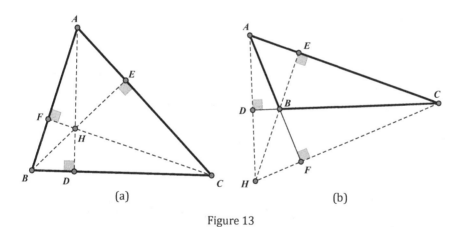

(a)   (b)

Figure 13

Curiosity 14. A Circle Intersects a Triangle to Generate Lots of Equal Segments

When circles are combined with a triangle, many unexpected results can appear. This time, we consider a triangle *MNR* formed by joining the midpoints of the sides of triangle *ABC* as shown in Figure 14, where point *H* is the orthocenter formed by the altitudes *AD*, *BE*, and *CF*. A random circle is drawn with center *H*, which intersects the sides *NR*, *RM*, and *MN* (extended) of triangle *MNR* at points (*K, L*), (*P, Q*), and (*S, T*), respectively. Quite surprisingly, we find lots of equality: *AK = AL = BP = BQ = CS = CT*.

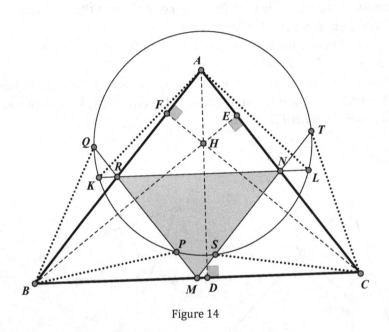

Figure 14

Curiosity 15. A Strange Appearance of a Congruent Triangle

At first glance, the configuration in Figure 15 appears to be somewhat complicated. However, it is actually a simple construction – when taken in small parts – that reveals an astonishing result. We begin with triangle *ABC* and the bisectors of each of its exterior angles, *DE*, *EF*, and *FD*, marking their points of intersections as *D*, *E*, and *F* as shown. We then locate the orthocenters of each of the three triangles, *BCD*,

CAE, and *ABF* and mark them as *G*, *H*, and *K*, respectively. What results is a triangle *GHK* whose sides are parallel and equal to the original triangle *ABC*. A bit lengthy, but certainly worth the amazing result – a pair of congruent triangles, namely, *GHK* and *ABC*.

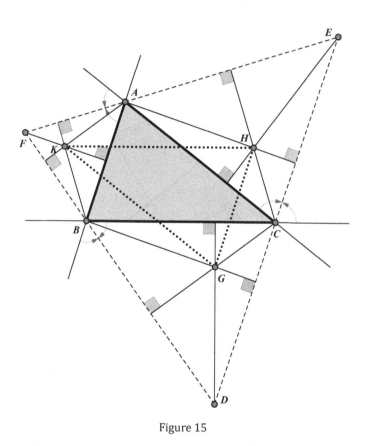

Figure 15

Curiosity 16. The Orthocenter Joins Three Other Points in Collinearity

Once again, we find that the altitudes of a triangle lead to some astonishing experiences. In Figure 16, point *P* is any point on the circumscribed circle of triangle *ABC*. We also have extended the altitudes *AL*,

BM, and *CN* of triangle *ABC* beyond the side to which they are drawn to intersect the circumscribed circle at points *D*, *E*, and *F*. When we construct the lines from point *P* to these three points *D*, *E*, and *F* on the circle, they intersect the corresponding sides of the triangle at points *R*, *S*, and *T*. Amazingly, these three points, *R*, *S*, and *T*, are collinear along with the orthocenter, *H*, of the triangle.

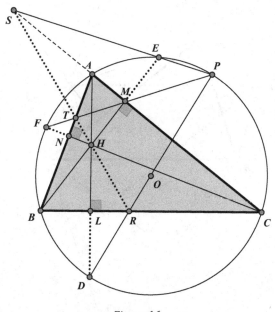

Figure 16

Curiosity 17. The Orthocenter Appears as the Midpoint of a Line Segment

The point of intersection of the altitudes, which is the orthocenter *H* of triangle *ABC*, is joined to the midpoint *M* of side *BC*, as shown in Figure 17. The line perpendicular to *MH* at point *H* intersects sides *AB* and *AC* at points *K* and *L*, respectively. Astoundingly, we find that *H* is the midpoint of line segment *KL*, that is, *KH* = *LH*.

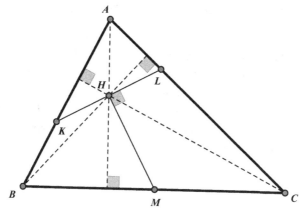

Figure 17

Curiosity 18. A Conglomeration of Perpendiculars Generates Equal Line Segments

We begin the configuration shown in Figure 18 with point H as the orthocenter of triangle ABC, whose altitudes are AD, BE, and CF. We then draw the line segment AP perpendicular to EF at point P. If we then also draw the line segment PD, the perpendicular line from H to EF intersects EF at point Q, and when QH is extended, it intersects PD at point R. Surprisingly, we find that $HR = HQ$.

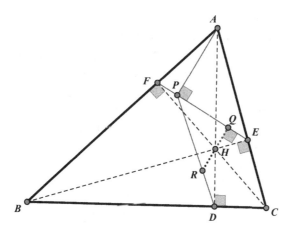

Figure 18

Curiosity 19. Yet Another Unexpected Collinearity

In Figure 19, a circle with center I is inscribed in triangle ABC. Points D, E, and F are its points of tangency with sides BC, CA, and AB, respectively. When the lines joining these points of tangency intersect the opposite sides (extended), we find that the resulting points of intersection are collinear. Here, EF intersects BC at point K, FD intersects CA at point L, and DE intersects AB at point M. Surprisingly, the points M, L, and K are collinear.

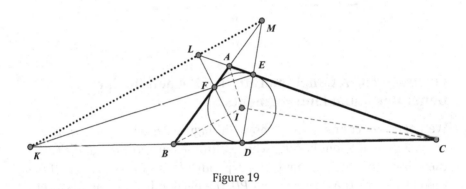

Figure 19

Curiosity 20. Introducing the Orthic Triangle

The feet of the altitudes of a triangle produce a triangle that is called an *orthic triangle*, which is a special type of *pedal triangle*,[1] and which has some very interesting properties. In Figure 20, we show one such feature, where we find that the altitudes of (acute) triangle ABC bisect the angles of the orthic triangle DEF. Furthermore, there are other features for us to appreciate in this configuration. One such is that there are four similar triangles to be seen here: $\triangle ABC \sim \triangle BFD \sim \triangle ECD \sim \triangle EFA$

---

[1]A pedal triangle is a triangle whose vertices are the feet of perpendiculars emanating from a common point in a triangle to the sides of a triangle.

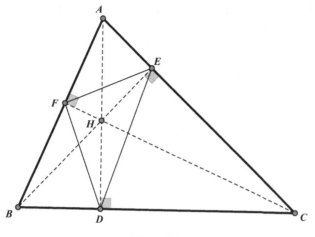

Figure 20

Curiosity 21. Finding the Perimeter of the Orthic Triangle

There is a rather nifty way of finding the perimeter of the orthic triangle *DEF*. When we draw the perpendiculars from point *D* to *AB* and from point *D* to *AC* meeting the sides at points *Q* and *P*, respectively, as can be seen in Figure 21, quite astonishingly, we find that the perimeter of triangle *DEF* is equal to twice the length of *PQ*.

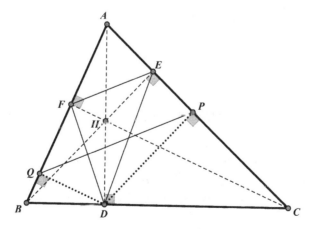

Figure 21

Curiosity 22. The Orthic Triangle is a Triangle's Smallest Inscribed Triangle

Another unforeseen feature of the orthic triangle is that it has the smallest perimeter of any other triangle inscribed in the same triangle. In Figure 22, we have orthic triangle *DEF* inscribed in triangle *ABC*. We then draw any other randomly inscribed triangle, in this case, triangle *JKL*, and we can show that the perimeter of the orthic triangle *DEF* is less than the perimeter of triangle *JKL*. This was initially discovered in 1775 by Giovanni Fagnano (1715–1797), and is therefore, referred to as *Fagnano's problem.*

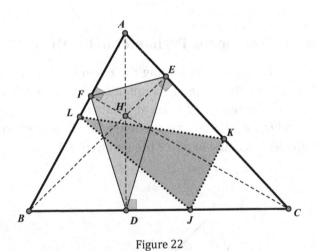

Figure 22

Curiosity 23. The Orthic Triangle's Surprising Similar Partner

There are times when we stumble on a relationship of similarity in a most unusual fashion. In Figure 23, we extend each of the altitudes of triangle *ABC* to meet the circumscribed circle at points *K, L,* and *M*. The resulting triangle *KLM* has sides parallel to those of the orthic triangle *DEF*, which results in the triangle similarity $\triangle KLM \sim \triangle DEF$.

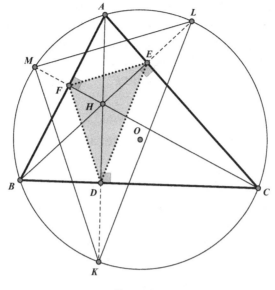

Figure 23

Curiosity 24. A Nice Concurrency Generated by the Orthic Triangle

The orthic triangle continues to amaze us with further concurrencies. We consider triangle *ABC*, shown in Figure 24, along with its orthic

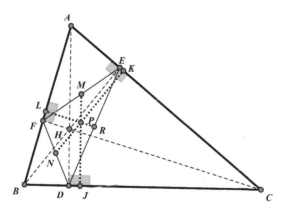

Figure 24

triangle *DEF* and the midpoints *M*, *N*, and *R* of sides *EF*, *FD*, and *DE*, respectively. From these midpoints, perpendiculars *JM*, *KN*, and *LR* are drawn to the opposite sides of the original triangle *ABC* meeting the sides at points *J*, *K*, and *L*. Once again, we find an unanticipated concurrency, as *JM*, *KN*, and *LR* meet at point *P*.

Curiosity 25. Orthic Triangle Generates an Isosceles Triangle

The orthic triangle offers yet another curiosity, as it will help to produce an isosceles triangle. Suppose we take a midpoint on any one of the sides of triangle *ABC*, as shown in Figure 25. We choose the midpoint *M* of side *AC*. When we join point *M* with the two opposite vertices of the orthic triangle *DEF*, quite unexpectedly, we find that the resulting segments are equal, so that we have *DM* = *FM*, thus making triangle *DMF* isosceles. We can then form other isosceles triangles by simply selecting the other midpoints on the sides of the triangle and connecting these points with the respective opposite vertices of the orthic triangle.

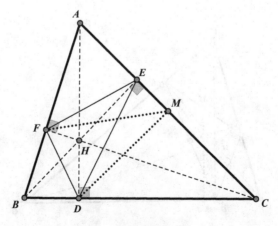

Figure 25

Curiosity 26. The Orthic Triangle Generates a Parallelogram

As we have seen from the previous curiosities, there seem to be endless amazing properties that evolve from the orthic triangle. As we will now see, the orthic triangle can produce an unexpected parallelogram. If we consider the inscribed circle of orthic triangle *DEF*, shown in Figure 26, with points of tangency *K*, *L*, and *M*, and extend *KL* to meet altitude *AD* at point *G*, surprisingly, we have *GL* = *GM*. That's not all. When we extend *LM* to meet altitude *FC* at point *J* and then draw *JE* and *EG*, we find that we have now created a parallelogram *JLGE*.

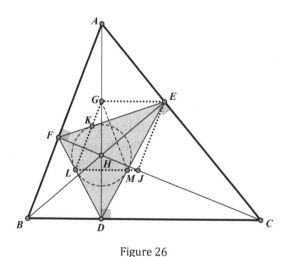

Figure 26

Curiosity 27. Concyclic Points Generated by the Orthic Triangle

If that isn't enough, we still have more surprises in this configuration, as we can see in Figure 27. Here we have six concyclic points: *H*, *M*, *J*, *E*, *G*, and *K*. All of these points lie on a common circle that has its center *O* on altitude *BE*. Having six points on one circle is quite a rarity.

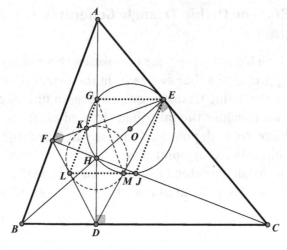

Figure 27

Curiosity 28. The Orthic Triangle Generates Concurrent Lines

There are times when collinearity appears under the strangest circumstances, and once again, the orthic triangle will take us down this path. In Figure 28, we have the orthic triangle *DEF* in triangle *ABC*, where *AD*, *BE*, and *CE* are its altitudes. We select a point *P* on altitude *AD*. The line *EP* is extended to meet *FD* at point *G*, and the line *FP* is

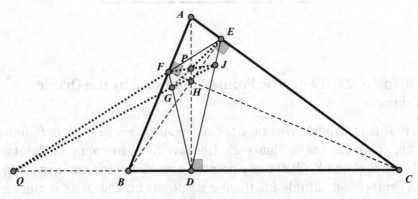

Figure 28

extended to meet the line *ED* at point *J*. The unexpected result is that *EF*, *JG*, and *CB* when extended all meet at a common point *Q*. Here we can see how somewhat convoluted geometry leads to a delightful result.

Curiosity 29. The Orthic Triangle Generates Collinear Points

In a similar configuration as that in Figure 28, we will now consider an extension, which is shown in Figure 29, where we are led to a collinearity instead of a concurrency. Here, the feet of the altitudes to the sides of triangle *ABC* are at points *D*, *E*, and *F*. The line *DE* intersects *AB* (extended) at point at *M*; the line *EF* intersects side *BC* (extended) at point *K*; and *FD* intersects side *CA* (extended) at point *L*. Unexpectedly, we then find that the points *K*, *L*, and *M* are collinear.

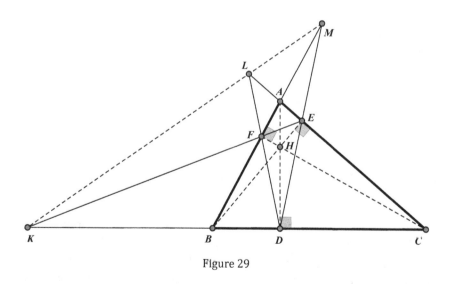

Figure 29

Curiosity 30. An Unexpected Property of Altitude Feet of a Triangle

In Figure 30, we see a general triangle *ABC*, with altitudes *BE* and *CF* meeting sides *AC* and *AB* at points *E* and *F*, respectively. From the midpoint *M* of side *BC*, we find that *ME* = *MF*, as we have already established in Curiosity 25. What makes this so unusual is that it is true for any shape triangle *ABC*. The configuration also has a few more surprises waiting for us. We find that ∠*EBA* =∠*ACF*, and after constructing *EF*, we also find that the perpendicular bisector of *EF* contains the midpoint *M* of side *BC*.

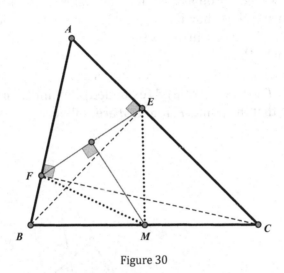

Figure 30

Curiosity 31. How a Non-Isosceles Triangle Can Generate a Parallelogram

Following various steps within an ordinary triangle – one that has no special properties – can frequently generate beautiful geometric figures. In Figure 31, for instance, we see how we can generate a parallelogram from a randomly drawn non-isosceles triangle *ABC*. The circumscribed circle of *ABC* has center point *O*. We draw *AO* extended to point *D*, so that ∠*ADB* = ∠*CDA*. From point *D* we draw a perpendicular to *AB* extended, intersecting it at point *E*, and a perpendicular to *AC* extended, intersecting it at point *F*. We need one more point to determine the parallelogram and that is the point *G*, which is the foot of the altitude from *A* to *BC*. We can now admire that the quadrilateral *DFGE* is, in fact, a parallelogram.

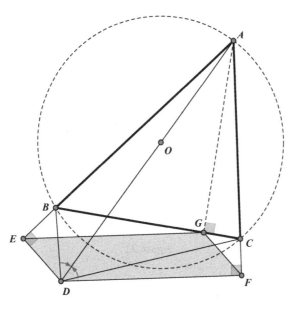

Figure 31

Curiosity 32. Perpendiculars Generating Unexpected Parallel Lines

Drawing perpendiculars to the sides of a triangle can have some highly unexpected results, as we can see in Figure 32. Perpendiculars are drawn from a randomly selected point *P* on the circumscribed circle of triangle *ABC* to each of the three sides of triangle *ABC*, intersecting sides *BC*, *CA*, and *AB* at points *J*, *K*, and *L*, respectively. These perpendiculars (some extended) then intersect the circumscribed circle a second time in points *D*, *E* and *F*, respectively. When we connect these to the appropriate vertices of the triangle, a series of three parallel lines evolve, namely, *AD*, *BE*, and *CF*. Remember, the position of point *P* could have been anywhere on the circumscribed circle, and these parallels would always result.

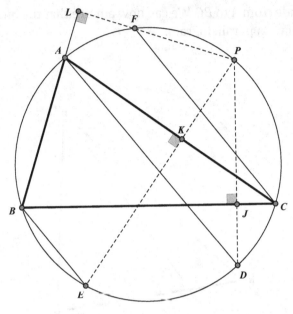

Figure 32

Curiosity 33. Inscribed Triangle Generates Collinearity

Collinearity also results when the three tangents to the circle circumscribed about a triangle at the vertices of the inscribed triangle meet the opposite sides of the triangle. We can see this in Figure 33, where tangents *AD*, *BE*, and *CF* to the circumscribed circle *O* of triangle *ABC* meet the opposite sides *BC*, *CA*, and *AB* at points *D*, *E*, and *F*, respectively, which turn out to be collinear.

Curiosity 34. Unexpected Congruent Triangles in a Circle

Creating two congruent triangles with different orientations inscribed in the same circle is truly an unusual experience. In Figure 34, we revisit the configuration from curiosity 32. Triangle *ABC* is inscribed in circle *O*, and from point *P*, which is any point on circle *O*, perpendiculars are drawn to the three sides, meeting them at points *J*, *K*, and *L*. These perpendiculars intersect the circle a second time in *D*, *E*, and *F*, respectively. Quite unexpectedly, we find that $\triangle ABC \cong \triangle DFE$.

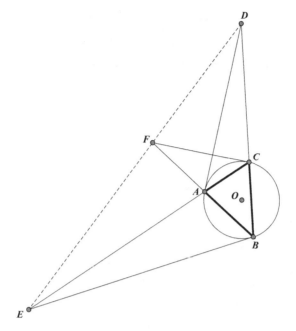

Figure 33

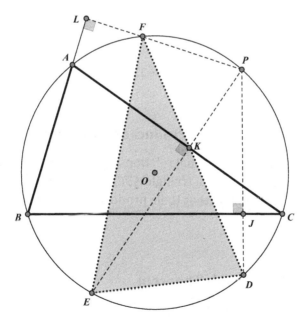

Figure 34

Curiosity 35. An Unexpected Product Equality

In Figure 35, we see triangle *ABC* with altitude *BD*, inscribed in circle *O* with diameter *BE*. Remarkably, this simple arrangement yields $AB \cdot BC = BD \cdot BE$.

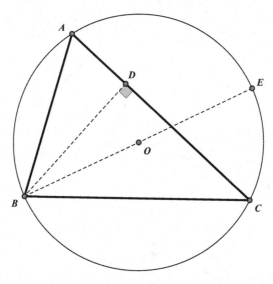

Figure 35

Curiosity 36. An Unusual Product of Two Sides of a Triangle

In Figure 36, we see an unusual relationship between the diameter of the circumscribed circle of a triangle *ABC*, its altitude *AE* and its sides *AB* and *AC*. It so happens that their products are equal, and we have $AD \cdot AE = AB \cdot AC$.

Curiosity 37. Simson's Theorem

A remarkable result concerning collinear points in triangles is the famous Simson theorem. This theorem involves one of the great

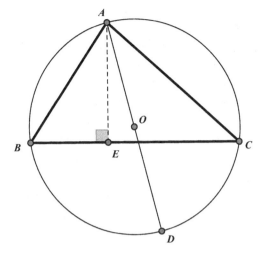

Figure 36

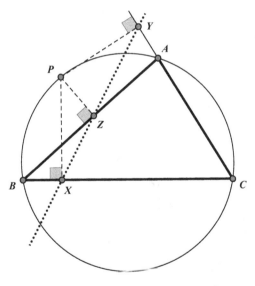

Figure 37

naming injustices in the history of mathematics. It was originally published by William Wallace (1768–1843) in Thomas Leybourn's *Mathematical Repository* (1799–1800). Through careless misquotes, the theorem was attributed to Robert Simson (1687–1768), the famous English interpreter of Euclid's *Elements*, whose book was the basis for the study of geometry in the English-speaking world and, more specifically, greatly influenced the American high school geometry course. To conform to the norm, we shall nevertheless stick to the popular reference *Simson's theorem*.

Simson's theorem states that the feet of the perpendiculars drawn from *any* point on the circumscribed circle of a triangle to the sides of the triangle are collinear. This is shown in Figure 37, where point P is any point on the circumscribed circle of triangle ABC. We then draw $PX \perp BC$, $PY \perp CA$, and $PZ \perp AB$, with X on BC, Y on CA, and Z on AB. According to Simson's theorem, points X, Y, and Z are collinear, regardless of where point P is positioned on the circumscribed circle of the triangle. This line is usually referred to as the *Simson line*.

Curiosity 38. An Extension of Simson's Theorem

An interesting extension to Simson's theorem concerns the case where the point P lies on the extension of one of the altitudes of triangle ABC.

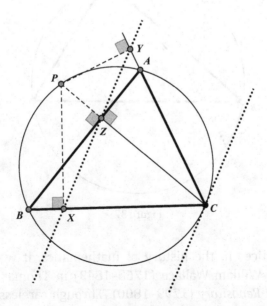

Figure 38

In the case shown in Figure 38, *P* lies on the altitude (extended) from vertex *C* to side *AB*. The result is that the Simson line is then parallel to the tangent to the circumscribed circle at point *C*.

Curiosity 39. An Interesting Aspect of Simson's Theorem

The Simson line has yet another amazing property. We find that it bisects the line joining the Simson point *P* on the circumcircle of *ABC* and the orthocenter *H* of *ABC*. In other words, as we see in Figure 39, we have *RP* = *RH*, where *R* denotes the intersection of *PH* and the Simson line of *P*.

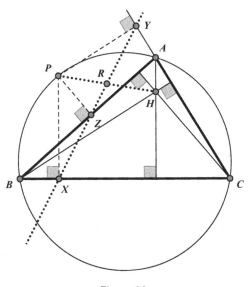

Figure 39

Curiosity 40. A Parallel to the Simson Line

When we extend one of the perpendiculars from point *P*, say perpendicular *PX*, as in Figure 40, to meet the circumscribed circle at point *Q*, lo and behold, we find that *AQ* is parallel to the Simson line of *P*.

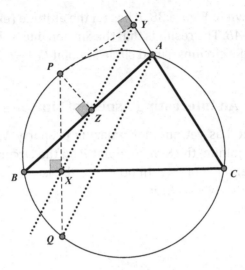

Figure 40

Curiosity 41. Two Triangles Related by a Common Point: Circumcenter – Centroid

Although it may be a bit cumbersome, there are times when two triangles are related by sharing a common point of concurrency; in this case, the center of the circumscribed circle of one triangle is the centroid of the other triangle. In Figure 41, the center of the circumscribed circle of triangle *ABC* is point *O*. We draw the medians of triangle *ABC*, which are *AL*, *BM*, and *CN* and meet at point *G*, the centroid of *ABC*. Constructing the perpendicular bisectors of *AG*, *BG*, and *CG*, which are at points *Q*, *R*, and *S* respectively, we find that these perpendicular bisectors meet at points *X*, *Y*, and *Z*. If we then draw the medians of triangle *XYZ*, namely, *XD*, *YE*, and *ZF*, they have their point of intersection (the centroid of *XYZ*) at point *O*. Therefore, point *O* is the centroid of triangle *XYZ* and the center of the circumscribed circle of triangle *ABC*.

Curiosity 42. Introducing the Medians of a Triangle with Some of Their Amazing Properties

Recall that the median of a triangle joins the triangle's vertex with the midpoint of the opposite side and the three medians meet at a

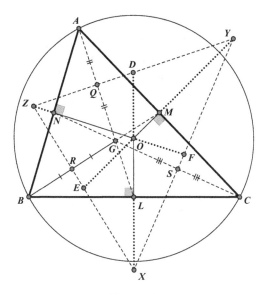

Figure 41

common point called the *centroid*, which is the center of gravity of the triangle. In Figure 42, we see triangle *ABC*, with medians *AD*, *BE*, and *CF* and centroid *G*. Any one of the three medians of a triangle divides the triangle into two equal area sections. For example, the median *AD* of triangle *ABC* partitions the triangle into two equal area triangles, and we have area [*ABD*] = area [*ACD*], since they have equal bases and share a common altitude. We can take this a step further to show that the medians of the triangle partition the entire triangle into six equal area triangles: area [*GBD*] = area [*GCD*] = area [*GCE*] = area [*GAE*] = area [*GAF*] = area [*GBF*].

This also implies the well-known fact that the centroid is a trisection point of each median.

The shortest median of a triangle is the one whose endpoint is on the longest side of the triangle. Furthermore, the sum of the lengths of the medians is less than the perimeter of the triangle, and greater than three-quarters of the perimeter the triangle. Referring to Figure 42, this would appear symbolically as:

$$AB + BC + CA > AD + BE + CF > \frac{3}{4}(AB + BC + CA)$$

The medians also have a rather unexpected equality relationship to the sides of the triangle, namely, the sum of the squares of the medians is equal to three-quarters of the sum of the squares of the sides of the triangle. In Figure 42, this means:

$$AD^2 + BE^2 + CF^2 = \frac{3}{4}\left(AB^2 + BC^2 + CA^2\right).$$

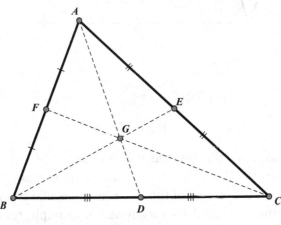

Figure 42

There are many more relationships involving the medians and the sides of a triangle worth pursuing.

Curiosity 43. The Median of a Triangle is Equidistant from Two Vertices

A not-well-known property of the median of a triangle is that it is equidistant from the other two vertices of a triangle. We notice that in Figure 43, in triangle *ABC*, median *AD* is extended to meet the perpendicular from vertex *B* to *AD* extended at point *P*. Also, the perpendicular from *C* to *AD* intersects it at point *Q*. We can then see that *BP* = *CQ*, which, in turn, indicates that the vertices at *B* and *C* are equidistant from the median *AD*.

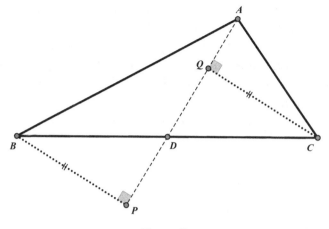

Figure 43

Curiosity 44. The Special Median of a Right Triangle

As might be expected, the right triangle has a special median. There, the median to the hypotenuse of a right triangle has half the length of the hypotenuse. In Figure 44, we have a right triangle *ABC*, and find that median $AM = \frac{1}{2}BC$.

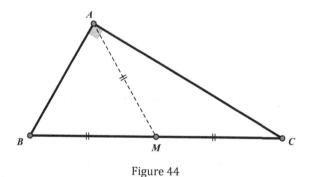

Figure 44

Curiosity 45. Medians Partition Any Triangle into Four Congruent Triangles

Besides partitioning a triangle into six equal area triangles, the medians of a triangle can also partition the triangle into four congruent triangles. If we join the midpoints of the sides of a triangle, as we have done in Figure 45, we then have line segments *DE*, *EF*, and *FD*. These are called the *midlines* of the triangle. They are bisected by the medians and the medians are in turn bisected by the midlines, so that for example, *AL* = *LD* and *EL* = *LF*. This is, of course, also true for the other medians. Furthermore, these midlines partition the triangle into four congruent triangles: Δ*AEF* ≅ Δ*FDB* ≅ Δ*ECD* ≅ Δ*DFE*

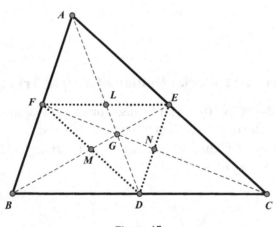

Figure 45

Curiosity 46. How the Centroid Helps Create a Similar Triangle

Aside from partitioning each median at a trisection point, as was mentioned in Curiosity 42, the centroid also allows us to construct a triangle similar to the original one, as we see in Figure 46. Here, the medians *AD*, *BE*, and *CF* determine the centroid *G*, and line *LK* is drawn through point *G* parallel to side *BC*, intersecting *AB* at point *L* and *AC* at point *K*. We then draw line *CL* to meet median *BE* at point *J* and draw line *BK* to meet median *CF* at point *H*. Surprisingly, we find that triangle *DHJ* is similar to triangle *ABC*.

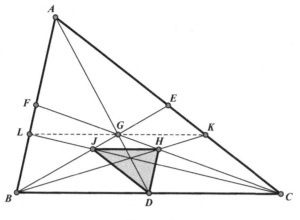

Figure 46

Curiosity 47. The Centroid can Provide a Most Unusual Balance

The centroid of the triangle has a most spectacular property. If you draw any line through the centroid, the sum of the distances from two vertices of the triangle to the line is equal to the distance from the third vertex to that same line. We see this in Figure 47, where the line

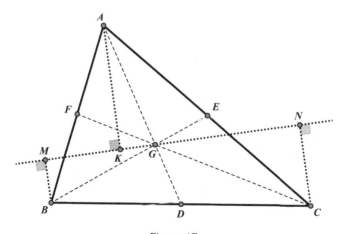

Figure 47

MN is *any* line through the centroid *G*, of triangle *ABC*. To measure the distance from the vertices of the triangle *ABC* to this line, we draw perpendiculars from each of the vertices to the line, namely, *AK*, *BM*, and *CN*, and we find that *AK* = *MB* + *NC*. Keep in mind that this is true for *any* line drawn through the centroid.

Curiosity 48. Distances from a Triangle's Vertices to a Random Line

Quite unexpectedly, we find a rather unusual relationship between the distances from the vertices of a triangle to a randomly drawn line and the distance from the centroid to that line. Consider Figure 48, where the distances from the vertices of triangle *ABC*, namely, *AK*, *BM*, and *CN*, are related to the distance from the centroid *G* to that same line, namely, *GP*. Curiously, they are related by *BM* + *AK* + *CN* = 3*PG*. It should be noted that the line *KMN* could have also passed through the triangle without containing the centroid. (In this case, the distances must be measured using "direction", so that points on one side of the line would have positive distances from the line, and points on the other side would have negative distances.)

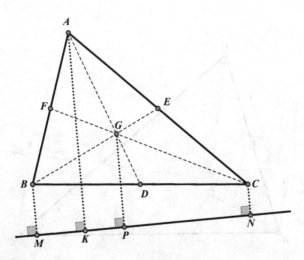

Figure 48

Curiosity 49. A Special Centroid Property When Two Medians are Perpendicular

Let us consider a triangle *ABC* with a peculiar characteristic, namely, that two of its medians are perpendicular. In Figure 49, triangle *ABC* has medians *AD* and *BE* that are perpendicular at the centroid point *G*. In this unusual case, the following interesting relationship holds true: $AC^2 + BC^2 = 5AB^2$.

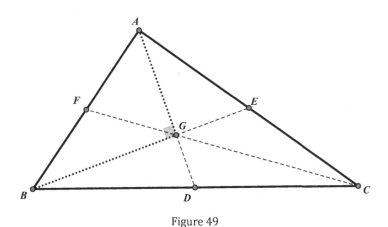

Figure 49

Curiosity 50. The Centroid's Amazing Property

The centroid also has a most unusual property that is not very well known, yet it is quite astonishing. In Figure 50, we have point *G* as the centroid of triangle *ABC*. The points *P* and *N* are randomly selected in triangle *ABC* with one condition: they are to be equidistant from the centroid *G*. The result, which usually baffles most, is that the sum of the squares of the distances from each of these two points to the three vertices of the triangle are equal. Symbolically, this can be stated as $AP^2 + BP^2 + CP^2 = AN^2 + BN^2 + CN^2$.

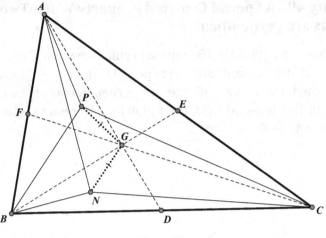

Figure 50

Curiosity 51. Some Median Surprises

The medians of a triangle also provide a number of other astonishing results. Consider, once again, triangle *ABC* with the medians *AD*, *BE*, and *CF*. We then draw *AJ* parallel to *BE*, and *BJ* parallel to *AC*, as shown in Figure 51. The first surprise is that the three points *J*, *E*, and *F* are collinear, and the second surprise is that *JC* bisects *FD*, so that *DN* = *FN*. These are some of the well-kept secrets in geometry!

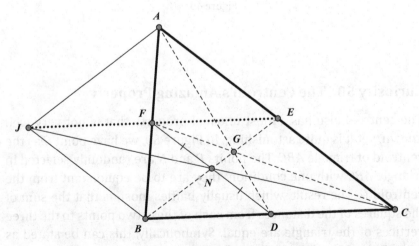

Figure 51

Curiosity 52. More Median Marvels

The medians of a triangle can sometimes offer some astonishing results as we can see in Figure 52. Here, we begin with triangle *AEF* and its three medians *AH*, *EJ*, and *FK*. When we extend *AH* its own length to point *C*, extend *FE* its own length to point *B*, and extend *EF* its own length to point *D*, the unexpected result is that the perimeter of triangle *ABC* is equal to the perimeter of triangle *ADC*, with both perimeters equal to twice the sum of the medians of triangle *ABC*. We can write this symbolically as: $AB + BC + CA = AD + DC + CA = 2(AH + EJ + FK)$.

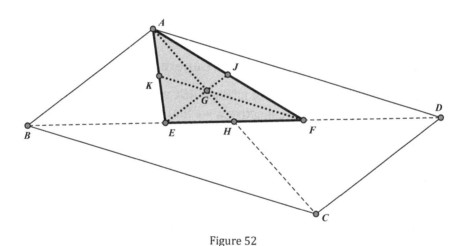

Figure 52

Curiosity 53. Comparing Medians to Triangle Perimeters

The medians of a triangle can also lead to some further curious results. We have the medians *AD*, *BE*, and *CF* of triangle *ABC*, meeting at point *G*. If we extend *DE* its own length to point *J*, we have created triangle *FCJ*, as shown in Figure 53. With a clever eye, we can see that the sum of the medians is equal to the perimeter of triangle *FCJ*. This can be expressed symbolically as $AD + BE + CF = FC + CJ + JF.$

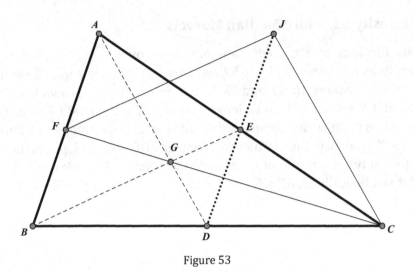

Figure 53

Curiosity 54. Median Extensions Generate Collinearity

When the medians BE and CF of triangle ABC are extended their own lengths to points P and Q, respectively, we find that the points P and Q are collinear with vertex A, as shown in Figure 54.

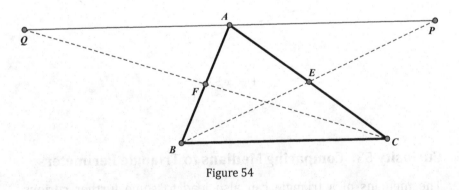

Figure 54

Curiosity 55. Two Unusual Triangles Share a Common Centroid

When one triangle is inscribed in another triangle, as we can see in Figure 55 with triangle QRS inscribed in triangle ABC so that the

points of contact determine the following proportion: $\dfrac{AS}{SB} = \dfrac{BQ}{QC} = \dfrac{CR}{RA}$,
the two triangles, QRS and ABC share a common centroid point. In Figure 55, the intersection of the medians QH, RJ, and SK (the centroid) of triangle QRS is at point G. This is also the intersection point for the medians of triangle ABC, namely, AD, BE, and CF.

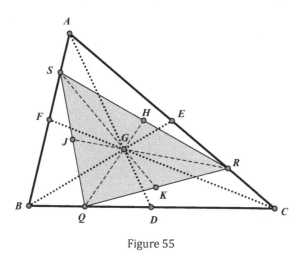

Figure 55

Curiosity 56. The Circumscribed Circle Revisited: The Incredible Relationship of the Centers of the Circumscribed Circles of the Median Triangles

Recall that the circumscribed circle of a triangle has its center at the point of intersection of the perpendicular bisectors of each of the sides, which we see in Figure 56(a). The perpendicular bisectors OD, OE, and OF meet at point O, which is the same distance from each of the vertices of triangle ABC. The circle with center O through A, therefore, also goes through B and C. We recall that the point O is called the *circumcenter* of ABC, while the circle is referred to as the *circumcircle*.

Having now identified the center of the circumscribed circle of a triangle, we can present a truly astonishing concurrency. Recall that the three medians divide the triangle ABC into six triangles of equal

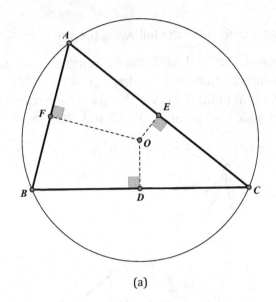

(a)

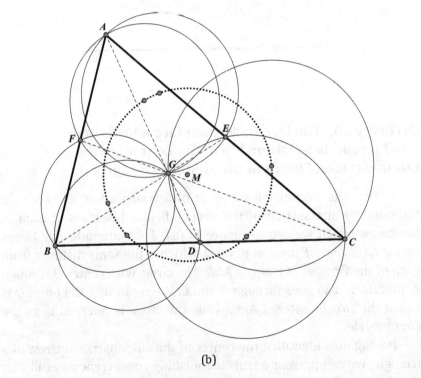

(b)

Figure 56

area. The circumscribed circles of each of these six equal-area triangles all share the common point *G*, as we can see in Figure 56(b). Surprisingly, the centers of these six circumscribed circles all lie on a common circle with center *M*.

Curiosity 57. An Astonishing Equality

In Figure 57, the line *AM* is the median to side *BC* of triangle *ABC*. Circle *O*, which contains vertex *A*, and is tangent to side *BC* at vertex *B*, intersects *AM* and *AC* at points *D* and *E*, respectively. The line through point *C* parallel to *BE* intersects *BD* extended at point *F*. Line *FE* extended meets *CB* extended at point *G*. Quite unexpectedly, we find that *AG = DG*.

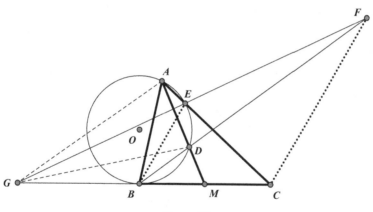

Figure 57

Curiosity 58. An Unexpected Simultaneous Bisection and Quadrisection in a Triangle

From simple constructions, surprising results can evolve. In Figure 58, we have a randomly drawn triangle *ABC*. Point *N* is selected on side *AB* so that *BN = 2AN*. We then select the midpoint *M* of side *BC* and draw *AM* and *CN*, which intersect at point *D*. Two curiosities evolve: point *D* is the midpoint of *AM*, and *DC = 3ND*. In other words, *D* bisects *AM* and quadrisects *CN*, since *DN* is one-quarter of *CN*.

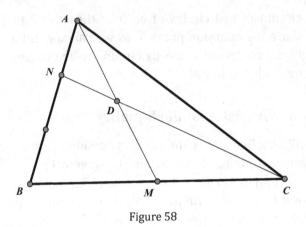

Figure 58

Curiosity 59. Noteworthy Triangle Area Relations

Surprising geometric results seem to be limitless. Consider the randomly drawn triangle *ABC*, shown in Figure 59. The midpoint *M* of side *BC* is joined with a randomly selected point *D* inside triangle *ABC*. We then find a most surprising result about the areas within this triangle, namely, area[*ADC*] − area[*ABD*] = 2·area[*ADM*].

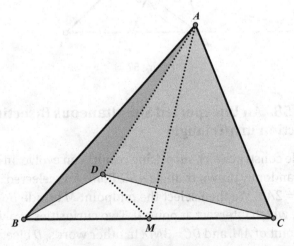

Figure 59

Curiosity 60. A Special Feature of a Random Point in an Equilateral Triangle

Now, let's investigate some unusual properties of the equilateral triangle. In Figure 60, we have drawn an equilateral triangle *ABC* and a random point within the triangle, which we call point *P*. The amazing fact about point *P* is that the sum of the distances from that randomly selected point to the sides of the equilateral triangle is the same as for any other point *P* inside the equilateral triangle that may have been selected. Thus, we have the following equation: $PD + PE + PF = AH$. The sum of these distances to the three sides is always equal to the altitude of the triangle, no matter which point we choose as *P*. This relationship is attributed to the Italian mathematician Vincenzo Viviani (1622–1703), who was a student of the famous scientist Galileo Galilei (1564–1642).

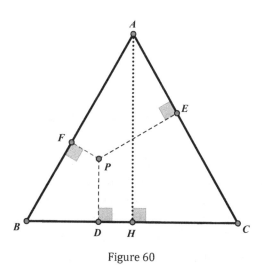

Figure 60

Curiosity 61. A Special Feature of a Point Outside of an Equilateral Triangle

Having spoken about a point inside the equilateral triangle, there is also a very interesting relationship with a point that is outside the equilateral triangle. More specifically, as we see in Figure 61, with a point on the circumscribed circle of the equilateral triangle *ABC*. The following relationship exists, regardless of where the point *P* is on the circle: *AP + CP = BP*.

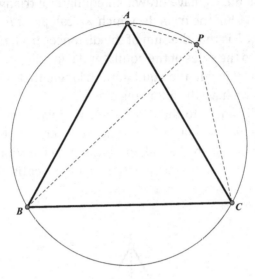

Figure 61

Curiosity 62. Using an Equilateral Triangle to Trisect a Line Segment

The unique features of the equilateral triangle also provide us with an opportunity to trisect a line segment. In Figure 62, angles *B* and *C* of equilateral triangle *ABC* are bisected by *PB* and *PC*, respectively. We then draw segments *PD* and *PE* parallel to the triangle sides *AB* and *AC*, respectively. The result is that the side *BC* has been trisected into equal parts at the points *D* and *E*, so that we have *BD = DE = EC*.

Curiosity 63. Another Way to Trisect a Line Segment with an Equilateral Triangle

There is yet another clever way to trisect the side of an equilateral triangle. As we see in Figure 63, point *M* is the midpoint of side *BC* of equilateral triangle *ABC*. External to triangle *ABC*, two equilateral

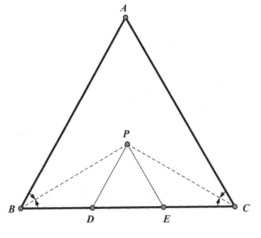

Figure 62

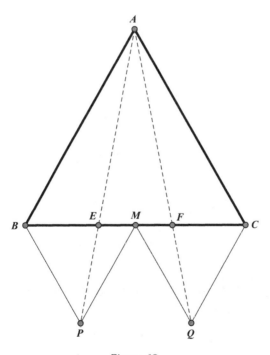

Figure 63

triangles are drawn, *BPM* and *MQC*. When we draw *AP* and *AQ*, they intersect *BC* at points *E* and *F*, so that *BE* = *EF* = *FC*. Lo and behold, once again the side of the equilateral triangle is trisected.

Curiosity 64. An Unusual Construction of a 30°-60°-90° Triangle Inside an Equilateral Triangle

In mathematics, a 30°-60°-90° triangle is often seen as very useful. The simplest way to create such a triangle is to draw an altitude in an equilateral triangle, whereupon two such triangles appear. However, in order to create such a triangle in a rather unusual way, we can proceed as follows. We begin with the equilateral triangle *ABC*. Point *D* is any point on side *AB* and line *DE* is parallel to *AC*, so that point *E* is on *BC*. Point *M* is the midpoint of *AE* and is connected to the center point *O* of equilateral triangle *DBE*. The resulting triangle *MOC* is then a 30°-60°-90° triangle, as shown in Figure 64.

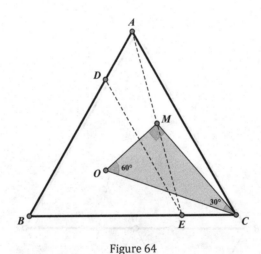

Figure 64

Curiosity 65. An Unusual Product of Segments in an Equilateral Triangle

In equilateral triangle *ABC*, a point *D* is selected randomly on the side *BC*, as shown in Figure 65. The line segment *EF* is the perpendicular bisector of *AD*, with points *E* and *F* on sides *AB* and *AC*, respectively. This results in the following relationship: *BE·FC = DB·DC*.

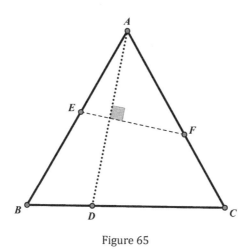

Figure 65

Curiosity 66. A Surprising Relationship Between an Isosceles Triangle and an Equilateral Triangle

Combining equilateral triangles and isosceles triangles can also produce some rather startling results. In Figure 66, we have an equilateral triangle *ABC* and an isosceles triangle *DBC*, so that $\angle BAC = 2\angle BDC$. The curious result is that two lines, which do not appear to be related in any way are, in fact, equal, as *AD = BC*.

Curiosity 67. An Unexpected Equality in an Isosceles Triangle

Figure 67 shows us a surprising result in an isosceles triangle *ABC*. Point *P* can take any position along the base *BC*. At the midpoints, *M* and *N*, of *BP* and *CP*, respectively, perpendiculars are erected to meet sides *AB* and *AC* at points *D* and *E*, respectively. Much to everyone's surprise, we find that *AE = DB* and *EC = AD*.

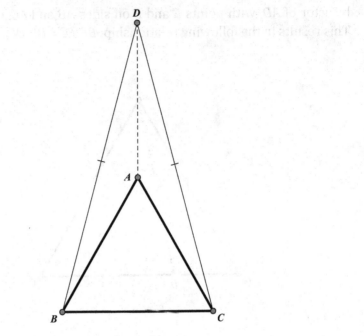

Figure 66

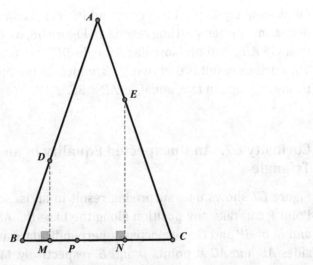

Figure 67

Curiosity 68. Another Unexpected Equality in an Isosceles Triangle

Once again, we begin with an isosceles triangle *ABC* with altitude *AM*, as shown in Figure 68. We extend side *BA* an unspecified length to point *P*. The perpendicular from *P* to side *BC* intersects side *AC* at point *R* and side *BC* at point *Q*. Quite unexpectedly, we find that *PQ* + *RQ* = 2*AM*.

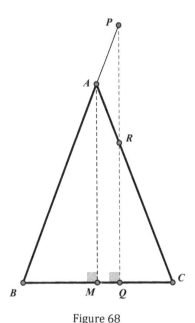

Figure 68

Curiosity 69. A Remarkable Geometric Equality

There are times when geometry offers us an almost inexplicable result. In Figure 69, we show an isosceles triangle *ABC* with *AB* = *AC*. A point *D* is selected anywhere on side *AB* and a point *E* is then determined on the extension of side *AC* such that *DE* is bisected by *BC* at point *F*. After having made this unusual construction, the surprising result is that we now have *DB* = *CE*.

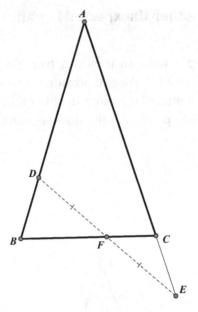

Figure 69

Curiosity 70. Another Remarkable Geometric Equality

A very simple construction leads to another quite surprising result. This time, in Figure 70, we begin with an isosceles triangle *ABC*, whose vertex angle at *A* is of measure 45°, and where *AD* is the altitude to base *BC*. From point *C*, a perpendicular intersects *AB* at point *E*. Finally, this perpendicular, *CE*, intersects *AD* at point *P*. Unexpectedly, we find that *PE* = *BE*.

Curiosity 71. An Unexpected Perpendicularity

Perpendicular lines seem to come up at the most unexpected times. In the configuration shown in Figure 71, *M* is the midpoint of the base *BC* of isosceles triangle *ABC*. A perpendicular line is constructed from point *M* to side *AC*, intersecting it at point *D*. Connecting the midpoint *F* of *MD* to the triangle vertex *A*, we find that *AF* turns out to be perpendicular to *BD* at point *E*.

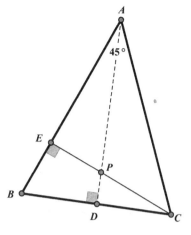

Figure 70

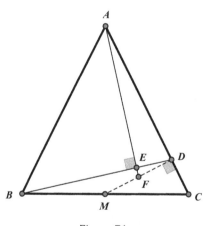

Figure 71

Curiosity 72. The Unexpected Property of an Altitude to the Side of an Isosceles Triangle

Consider the altitude *BD* of isosceles triangle *ABC*, where *AB* = *AC*, as shown in Figure 72. The amazing result is that ∠*BAD* = 2∠*CBD*.

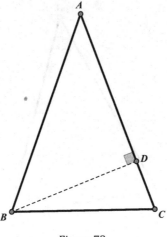

Figure 72

Curiosity 73. Peculiar Property of Isosceles Triangles

The isosceles triangle *ABC*, with *AB = AC*, holds another surprise in store for us. If we select any two points, *P* and *Q*, on base *BC*, as shown in Figure 73, much to our surprise, the sum of the distances to the two sides is the same. In other words, we have *PD + PE = QF + QG*. It should be emphasized that the points *P* and *Q* can be chosen anywhere along the base *BC*, and this will be true.

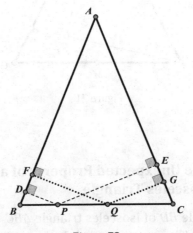

Figure 73

Curiosity 74. A Bizarre Connection: The Triangle Incenter on its Circumcircle

In Figure 74, a point *D* is placed randomly on side *AB* of isosceles triangle *ABC*, where *AB* = *AC*. It turns out that the center *I* of the inscribed circle of triangle *ADC* lies on the circumcircle of triangle *DBC*.

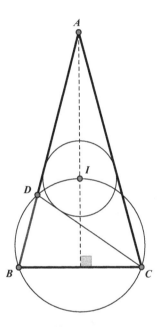

Figure 74

Curiosity 75. Collinear Points Generate an Angle Bisector

In triangle *ABC*, shown in Figure 75, a point *Q* is selected on the median *AM*. A perpendicular is drawn from *Q* to *BC*, intersecting it at point *D*. A point *P* is then chosen on *QD*, whereupon perpendiculars *PE* and *PF* are drawn to sides *AC* and *AB*, intersecting the sides at points *E* and *F*, respectively. As it happens, points *E*, *Q*, and *F* turn out to be collinear. In this case, we surprisingly find that *PA* bisects ∠*BAC*.

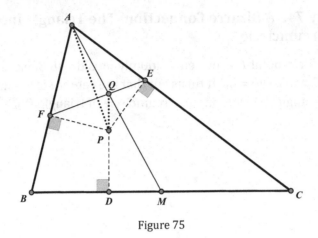

Figure 75

Curiosity 76. Unexpected Similar Triangles

Similar triangles can sometimes appear when we would least expect it. For example, in Figure 76, the interior angle and exterior angle at vertex *A* of triangle *ABC* are bisected by lines *AG* and *AH*, respectively, which intersect *BC* (extended) at points *G* and *H*, respectively. The midpoint of *GH* is point *M*. The result is that we have two triangles that are similar, namely, Δ*ABM* ~ Δ*ACM*.

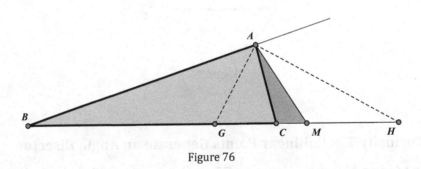

Figure 76

Curiosity 77. Perpendiculars to Four Angle Bisectors of a Triangle Reveal Four Collinear Points

An unexpected collinearity of four points can be seen when we draw angle bisectors to both interior angles and exterior angles, as we show in Figure 77. Triangle *ABC* has interior angle bisectors *CG* and *BD*, and exterior angle bisectors *BE* and *CJ*. When we draw perpendiculars from vertex *A* to each of these four angle bisectors, meeting them at points *G*, *D*, *E*, and *J*, respectively, we find that these four points lie on the same line, that is, they are collinear.

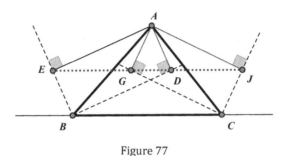

Figure 77

Curiosity 78. A Convoluted Bisection of a Line Segment

There are many simple ways to bisect a line segment. For entertainment, we will bisect a line segment in a rather convoluted fashion, but

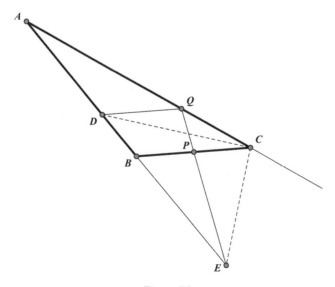

Figure 78

one that demonstrates the power of geometry. We begin with triangle *ABC*, shown in Figure 78, where we bisect the interior angle and exterior angle at vertex *C* with angle bisectors *CD* and *CE*. These intersect side *AB* (extended) in *D* and *E*, respectively. We then construct line *DQ* parallel to *BC* and intersecting *AC* at point *Q*. When we draw *QE*, we find that it bisects *BC* at point *P*, so that *BP* = *CP*.

Curiosity 79. Determining the Perimeter of a Triangle Without Measuring its Side Lengths

When we talk about amazing and truly unexpected results, we would probably consider the following as a prime example. We begin with any shape triangle *ABC*, as shown in Figure 79, with the sides *AB* and *AC* extended to *D* and *E*, respectively. We construct the bisectors *BG* and *CG* of the exterior angles ∠*DBC* and ∠*BCE*, respectively. These bisectors meet at point *G*. From point *G*, we construct a perpendicular

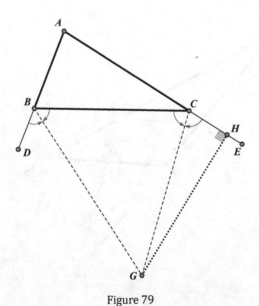

Figure 79

GH to line *AE* at point *H*. Here is the amazing result: the perimeter of triangle *ABC* is then equal to twice the length of *AH*.

Curiosity 80. A Strange Angle Trisection

For centuries, one of the most famous problems of geometry has been to figure out a way of trisecting an angle using only an unmarked straightedge and compasses. For the general angle, this is known to be impossible. However, we can use some clever geometry for trisecting angles under certain special circumstances, as we can see in Figure 80. We begin with triangle *ABC*, where *AB* is twice as long as *BC*. We first construct a perpendicular to *AB* at its midpoint *R*, and then construct a line perpendicular to *BC* at point *C*. These two perpendiculars intersect at point *P*. The unanticipated result is that *PR* trisects angle *APC*, so that $\angle APR = \frac{1}{3} \angle APC$. Thus, we have trisected an angle using merely an unmarked straightedge and compasses.

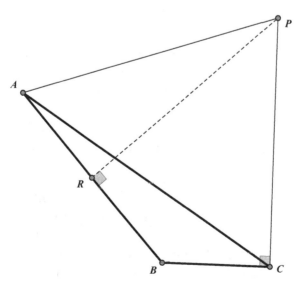

Figure 80

Curiosity 81. A Most Unexpected Equality

In Figure 81, we have chosen random points P and Q on sides AB and AC of triangle ABC, respectively. We have then drawn segments BX and CY as the bisectors of angles QBA and ACP, respectively. These bisectors intersect at point R. Remarkably, we then have $\angle BPC + \angle BQC = 2\angle BRC$.

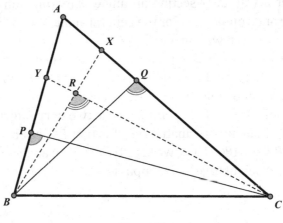

Figure 81

Curiosity 82. A Triangle Peculiarity

This triangle curiosity is certainly a peculiarity. When we determine the midpoints D, E, and F of the three sides BC, AC, and AB of triangle ABC, as shown in Figure 82, and construct the altitude CH from vertex C to side AB, we find that $\angle DFE = \angle DHE$.

Curiosity 83. A Counterintuitive Area Equality of Triangles

Counterintuitive results are not uncommon in geometry, especially when two triangles turn out to be of equal area despite this not visually appearing to be the case. We see an example of this in Figure 83. Here, triangle ABC has points F and D placed on side AC so that $AD = CF$. Through points D and F lines are drawn parallel to side AB, which intersect side BC at points E and G, respectively. Amazingly, it turns out that area[AED] = area[AGF]. "Looks" can certainly be deceiving!

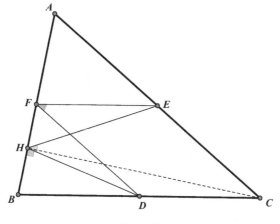

Figure 82

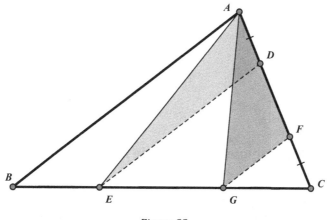

Figure 83

Curiosity 84. Parallel Lines Create a Double Area Triangle

Geometry often provides us with relationships that seem almost inexplicable. Using any shape triangle and drawing three parallel lines, we
can create a second triangle whose area is twice that of the first triangle. We show this in Figure 84, where we are given a random triangle
ABC and three parallels *AD*, *BE*, and *CF* through the vertices of the triangle, and intersecting the opposite sides (extended if necessary) at
points *D*, *E*, and *F*, respectively. The astonishing result is that the area
of triangle *DFE* is twice the area of triangle *ABC*.

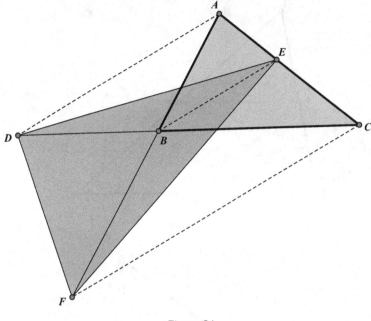

Figure 84

Curiosity 85. Unexpected Triangle Area Relationships

As we have seen from the previous examples, pictures illustrating area relationships in triangles can be quite deceiving. Consider a right triangle *ABC* with right angle at vertex *C*, as shown in Figure 85. An equilateral triangle *BDC* is drawn externally on side *BC*. Unexpectedly, we have the following relationship: area[*BDC*] + area[*ADC*] = area[*ABD*].

Curiosity 86. Creating a Triangle Whose Area is Three-Quarters the Area of a Given Triangle

Constructing a triangle whose area is three-quarters that of a given triangle can be a challenge. Yet, there is a very clever way that this can be done. We begin with triangle *ABC* as shown in Figure 86. We

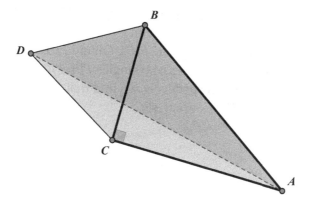

Figure 85

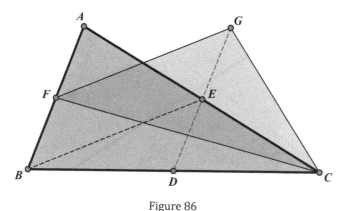

Figure 86

determine the midpoints of *BC*, *CA*, and *AB*, naming them *D*, *E*, and *F*, respectively. Next, we construct a line through *F* parallel to *BE*, meeting *DE* extended at point *G*. This results in a triangle *GFC* whose three sides have lengths equal to the medians of triangle *ABC*. Moreover, $\text{area}[GFC] = \frac{3}{4} \cdot \text{area}[ABC]$.

Curiosity 87. An Astounding Construction: Similar Triangles Whose Area Ratio is 1:4

In Figure 87, we see a configuration in which a triangle generates a second similar triangle that, curiously, has one quarter of the area of the original one. Choosing a random triangle *ABC* and selecting any point *P*, perpendiculars are drawn from *P* to each of the three sides of triangle *ABC*. We then have *PG* ⊥ *BC*, *PH* ⊥ *CA* and *PJ* ⊥ *AB* (extended). When we draw the circumcircles of triangles *PJH*, *PJG*, and *PHG*, with circle centers *K*, *L*, and *N*, respectively, we find that Δ*ABC* ~ Δ*KLN*, and furthermore, area[*KLN*] = $\frac{1}{4}$·area[*ABC*].

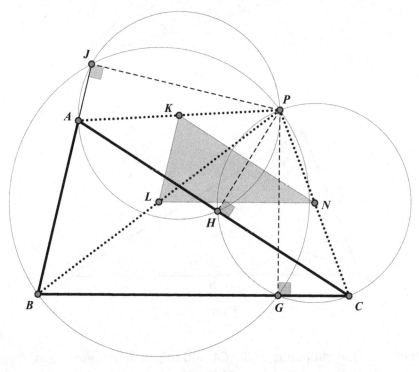

Figure 87

Curiosity 88. Unforeseen Equality of Inscribed Triangles

In Figure 88, triangle *DEF* and triangle *RST* are inscribed in triangle *ABC*, with each of their vertices on a side of the larger triangle. There is a most unexpected relationship that exists between two such triangles that are inscribed in the same triangle. When the vertices of each of the inscribed triangles are placed on each side of the larger triangle in such a way that their vertices are equidistant from the midpoint of the respective side, then the two inscribed triangles have equal areas. In Figure 88, K, M, and N are the midpoints of sides BC, CA, and AB, respectively. The two triangles DEF and RST inscribed in triangle ABC are placed such that $RK = DK$, $SM = EM$, and $TN = FN$. We can then conclude that area$[DEF]$ = area$[RST]$.

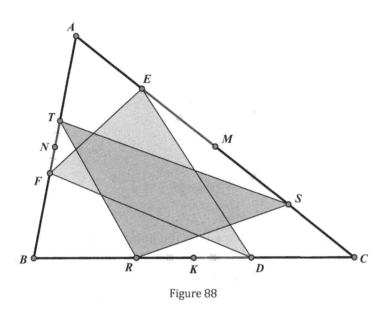

Figure 88

Curiosity 89. The Unanticipated Commonality of Equal Area Triangles

Triangles of equal area can also provide astonishing results. Consider two triangles $\triangle ABC$ and $\triangle BDC$ of equal area, sharing a common side BC, as we see in Figure 89. When we join the two remote vertices A and D, we find that the common side BC bisects the line AD, so that we have $AM = DM$.

Curiosity 90. How a Random Point Divides the Area of a Triangle in Half

It is surprisingly easy to determine a line through any point on the side of a triangle that partitions the triangle into two parts of equal

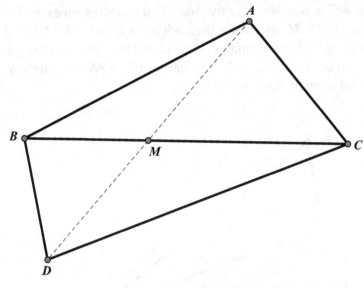

Figure 89

area. We see this in Figure 90, where the point *P* is at any position on side *AB* of triangle *ABC*, between the midpoint *M* of *AB* and vertex *A*. The point *N* on side *BC* is determined by the line parallel to *PC* emanating from point *M*. The amazing result is that the triangle *PBN* has one half the area of triangle *ABC*.

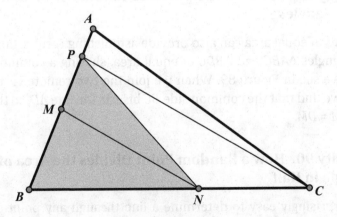

Figure 90

Curiosity 91. Determining a Triangle One-Third of the Area of a Given Triangle

Interestingly, a triangle with one-third the area of a given triangle results by trisecting each of the sides of the original triangle. Consider triangle *ABC*, where each of the sides is trisected as shown in Figure 91, where the trisection points are *D*, *E*, *F*, *G*, *H*, and *J*, with

$$AD = DE = EB, BF = FG = GC, \text{ and } CH = HJ = JA.$$

We then have area$\left[EGJ\right] = \frac{1}{3} \cdot$ area$\left[ABC\right]$. Furthermore, we also have

$$\text{area}[AEJ] = \text{area}[BGE] = \text{area}[CJG].$$

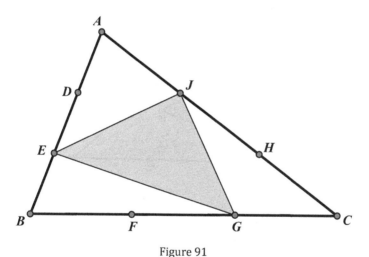

Figure 91

Curiosity 92. Trisection Points Partitioning a Triangle

In Figure 92, we again consider a triangle *ABC* with trisected sides. Points *D* and *E* are on sides *AB* and *AC*, respectively, such that *DB* = 2*DA* and *EC* = 2*EA*. Lines *BE* and *CD* intersect at point *F*. There are several noteworthy properties in this configuration. First, we note that triangle *BCF* has half the area of triangle *ABC*. Furthermore, we also have area[*ADFE*] = area[*BFD*] = area[*CEF*].

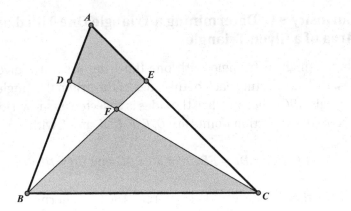

Figure 92

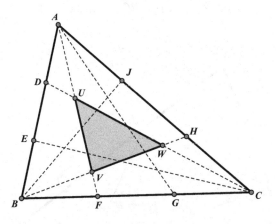

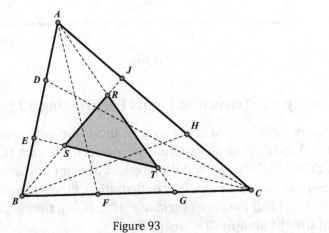

Figure 93

Curiosity 93. Further Surprises Provided by the Trisection Points of Triangle Sides

The trisection points of the sides of a triangle produce yet more amazing results. In Figure 93, each of the sides of triangle *ABC* is once again partitioned into three equal parts by points *D*, *E*, *F*, *G*, *H*, and *J*, with *AD = DE = EB*, *BF = FG = GC*, and *CH = HJ = JA*.

Lines are then drawn from each vertex to the trisection points on the opposite side. In the top part of Figure 93, intersection points have been selected to form triangle *UVW*, and it then turns out that area$[UVW]=\frac{1}{7}\cdot$area$[ABC]$. In comparison, in the bottom half, the points were selected to form triangle *RST* and we also find that area$[RST]=\frac{1}{7}\cdot$area$[ABC]$. This means that the two triangles *UVW* and *RST* have equal areas. It should be noted, however, that in general the two triangles are *not* congruent, *not* similar, and do *not* have the same perimeter.

Curiosity 94. Another Surprise Provided by the Trisection Points of Triangle Sides

In Figure 94, we see the same trisection points in triangle *ABC* as in Figure 93, along with the same connecting line segments. Triangle *XYZ*,

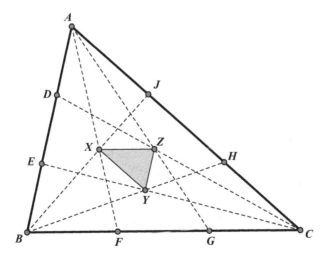

Figure 94

whose vertices are also intersections of these segments as shown, has the property area$[XYZ] = \frac{1}{25} \cdot$area$[ABC]$. Furthermore, we find that $\triangle XYZ$ is similar to $\triangle ABC$, as their corresponding sides are parallel, thus, making their corresponding angles equal.

Curiosity 95. Yet Another Surprise Provided by the Trisection Points of Triangle Sides

In Figure 95, we once again see the same trisection points in triangle *ABC* as in Figure 93, along with the same connecting line segments. This time, we consider triangle *LMN*, whose vertices are also intersections of these segments as shown. This triangle has the following property: area$[LMN] = \frac{1}{16} \cdot$area$[ABC]$. Triangle $\triangle LMN$ is similar to $\triangle ABC$, as their corresponding sides are once again parallel, thus, making their corresponding angles equal.

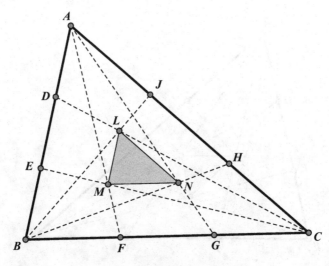

Figure 95

Curiosity 96. Medians and Trisectors Partitioning a Triangle with an Unexpected Result

More interesting triangle areas result when we add medians to the trisectors of a given triangle. In Figure 96, medians *AM* and *BN*, taken together with trisector *CD*, determine the triangle *LUZ*. Here we find that area$[LUZ] = \frac{1}{60} \cdot$area$[ABC]$. It should be noted that the triangle *LUZ* is in general not similar to triangle *ABC*.

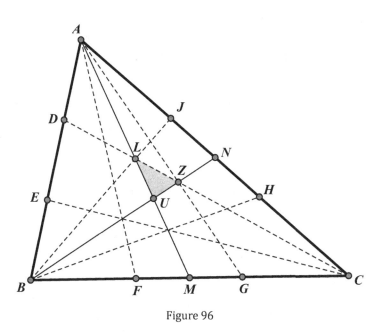

Figure 96

Curiosity 97. The Unforeseen Characteristic of a Random Triangle with a 60° Angle

It is curious that a randomly constructed triangle with one angle of measure 60° would have the unexpected property that its angle bisectors determine an equality of portions of two sides of a triangle compared to the third side. In Figure 97, triangle *ABC* has ∠*BAC* = 60°. Lines *AP* and *BQ* bisect angles *BAC* and *CBA*, respectively. Amazingly, we find that *BC* = *BQ* + *CP*.

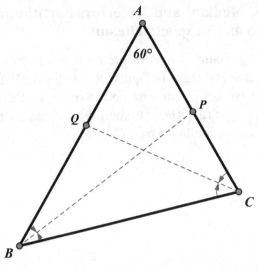

Figure 97

Curiosity 98. Another Surprising Feature of a Random Triangle with a 60° Angle

Once again, in Figure 98, we have drawn triangle *ABC* with ∠*A* = 60°. The altitudes *BE* and *CF* intersect at the orthocenter *H*. Quite surprisingly, we find that the line *OH*, joining the orthocenter with the center *O* of the circumscribed circle, bisects ∠*FHB*. As a further curiosity, we note that ∠*FHB* = 60°, as well.

Curiosity 99. Another Unexpected Collinearity

In Figure 99, we show triangle *ABC* with its inscribed circle center at *I* and its circumscribed circle center at point *O*. The line through *I* and parallel to side *AB* meets at point *P* the line tangent to the circumscribed circle at point *C*. The line *AI* (extended) intersects the circumscribed circle at point *D*, and the line *BI* (extended) intersects the circumscribed circle at point *E*. Unanticipatedly, we find that the points *D*, *E*, and *P* are collinear.

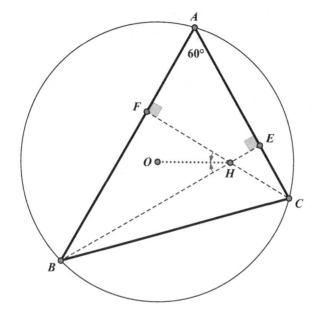

Figure 98

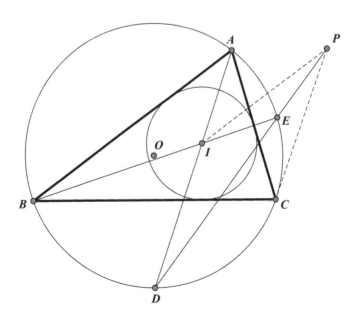

Figure 99

Curiosity 100. Reflections of Triangles Generate Concurrent Circles and Concurrent Lines

Concurrency of circles can also result in another rather beautiful relationship. Suppose we begin with any triangle *ABC* and reflect triangle *ABC* in each of its sides, so that the four triangles shown in Figure 100 are all congruent. When we draw a circumscribed circle about each of the three triangles $\triangle BA'C$, $\triangle CB'A$, and $\triangle AC'B$, we find that the three circumcircles share a common point *P*. We can even take this a step further and muse at another amazing group of concurrent lines, when we find that the lines *AQ*, *BR*, and *CS*, joining the centers *Q*, *R*, and *S* of the circles to the respective remote triangle vertices, all share a common point *M*.

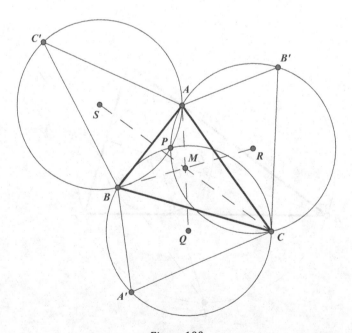

Figure 100

Curiosity 101. The Wonders of Three Concurrent Congruent Circles

A most unusual relationship exists when we have a given triangle and three concurrent congruent circles that are drawn in such a way that each circle contains two vertices of the triangle. We can see this configuration in Figure 101, where the three circles centered at Q, R, and S each contains two vertices of triangle ABC. The circles are concurrent in the point P. There are a number of notable features of this configuration.

(a) The circle containing points R, S, and Q also has the same radius as these original circles.

(b) The point of intersection P of the first three circles is the orthocenter (the intersection of the altitudes) of triangle ABC.

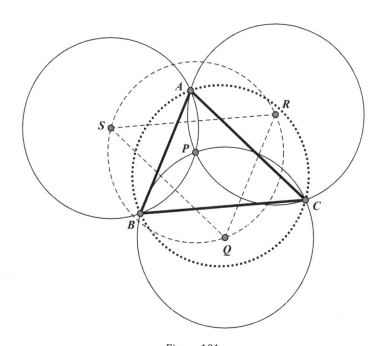

Figure 101

(c) The circumscribed circle of triangle *ABC* has the same radius as the three congruent circles.

(d) Triangle *QRS* is congruent to triangle *ABC*.

Curiosity 102. Further Unexpected Concurrencies

Concurrencies can often appear in the most unexpected ways. Suppose we select a random point *P* in the interior of triangle *ABC*, as illustrated in Figure 102. From that point we draw the perpendiculars to each of the sides, meeting the sides *AB*, *BC*, and *AC* at points *G*, *H*, and *J*, respectively. We then select a random point on each of the perpendiculars with point *D* on *PG*, point *E* on *PH*, and point *F* on *PJ*, thereby forming the triangle *DEF*. Next, we draw perpendiculars from the vertices of the original triangle *ABC* to each of the three sides of triangle *DEF*. We draw the perpendicular from point *A* to *DF*, the perpendicular from point *B* to *DE*, and the perpendicular point *C* to *EF*. Lo and behold, we find that these three perpendiculars are concurrent at a point *Q*. Bear in mind how randomly points were selected. Point *P* was any point in the triangle, and the triangle *DEF* had vertices placed randomly on the initial three perpendiculars. This is quite an unusual combination of concurrencies.

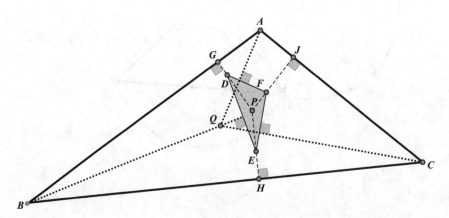

Figure 102

Curiosity 103. One Concurrency Generates Another Concurrency

The curiosity we show in Figure 103 is surely one of the most remarkable surprises regarding the concurrency of lines within triangles. We begin with triangle *ABC*, which has three concurrent line segments, *AD*, *BE*, and *CF*, intersecting at point *P*. When we draw a circle that contains points *D*, *E*, and *F*, we have the following amazing result: This circle intersects the sides of the triangle in three additional points, namely, points *X*, *Y*, and *Z*. The three segments joining these points with the opposite vertex, that is, *AX*, *BY*, and *CZ*, will always be concurrent as well. Bear in mind that our initial lines that intersected at point *P* could have been *any* three concurrent lines, and they then determined a circle which generated another concurrency at point *Q*. Truly amazing!

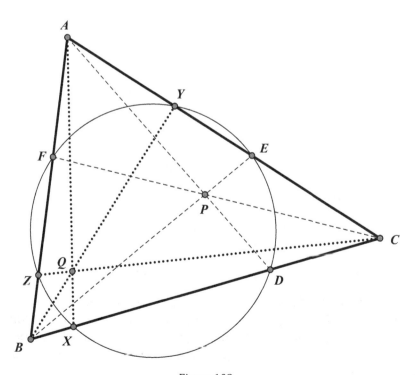

Figure 103

Curiosity 104. Unusual Perpendiculars Generating Concurrencies

There is another concurrency, which is analogous to the one we just considered in Figure 103. Here, we begin with a circle, which intersects each side twice of a randomly drawn triangle *ABC*, as shown in Figure 104. The perpendiculars *PD*, *PE*, and *PF*, drawn to each of the sides at three points of intersection with the circle, are known to be concurrent at point *P*. Amazingly, when perpendiculars are then drawn from the other three points of intersection *K*, *L*, and *M* with the circle, these perpendiculars *QK*, *QL*, and *QM* will unexpectedly also be concurrent at point *Q*. This is quite astonishing!

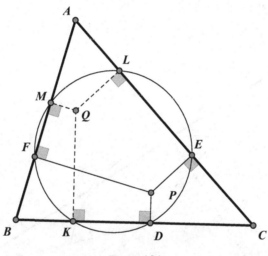

Figure 104

Curiosity 105. A Most Unexpected Concurrency

If the previous configuration was not impressive enough for you, the next one certainly ought to be. We begin in Figure 105 by drawing three concurrent lines *AD*, *BE*, and *CF* in triangle *ABC*, which intersect at point *P*. Next, we locate the midpoints *G*, *H*, and *J* of the three lines *AD*, *BE*, and *CF*, respectively. We then locate the midpoints *K*, *M*, and *N*

of the sides *BC*, *CA*, and *AB* of triangle *ABC*, respectively. Quite spectac-
ularly, when we join the previously determined midpoints to form the
segments *GK*, *HM*, and *JN*, we find that these lines are concurrent at a
point *Q*.

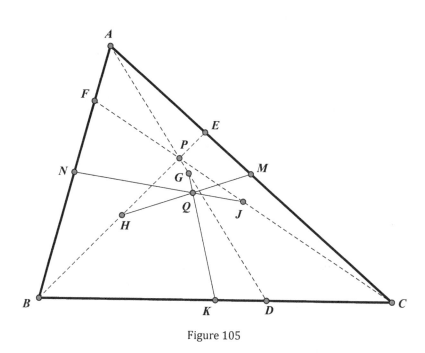

Figure 105

Curiosity 106. An Intriguing Concurrency

While we are on the hunt for more concurrencies in the triangle, here
is another one that is also quite unexpected, but true! This time, as
shown in Figure 106, we draw three concurrent lines *AD*, *DE*, and *CF*
in the triangle *ABC*, with the common point *P*. Using the feet *D*, *E*, and *F*
of these three line segments, we construct triangle *DEF* and locate the
midpoints *K*, *M*, and *N* of its sides *EF*, *FD*, and *DE*, respectively. Quite
amazingly, when the line segments *AK*, *BM*, and *CN* are extended, they
meet at a common point *Q*. Once again, a wonderful and unexpected
concurrency!

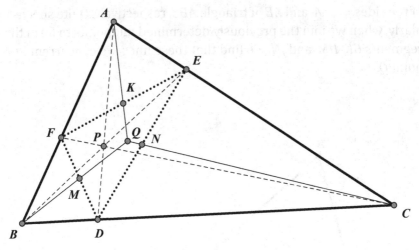

Figure 106

Curiosity 107. Another Unexpected and Unforeseen Concurrency

In Figure 107, we consider triangle *ABC* with altitudes *AP*, *BQ*, and *CR*. The points *D*, *E*, and *F* are the midpoints of sides *BC*, *AC*, and *AB*, respectively. The midpoints of segments *QR*, *PR*, and *PQ* are points *X*, *Y*, and *Z*, respectively. Truly unforeseen is the fact that *XD*, *YE*, and *ZF* are concurrent at a point *K*.

Curiosity 108. A Most Unexpected Concurrency from a Triangle

In Figure 108, we are again given a triangle *ABC*. Point *M* is the midpoint of side *AB*, and point *D* is any randomly selected point on side *AC*. The line *CE* is constructed through point *C* and parallel to side *AB*, and this line intersects line *BD* extended at point *E*. Side *BC* is then extended to intersect line *AE* at point *P*. Recalling that point *D* was randomly selected, we find that an unexpected concurrency evolves: The lines *AE*, *BC*, and *MD* (extended) are, in fact, concurrent at point *P*.

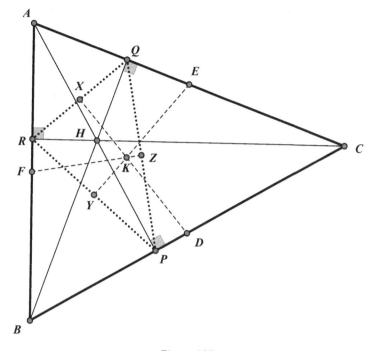

Figure 107

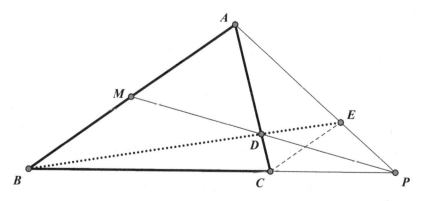

Figure 108

Curiosity 109. Astounding Point Property in a Triangle

Another surprising concurrency occurs in the configuration shown in Figure 109. Here, a random point P is selected in the interior of a triangle ABC with lines drawn from P to each of the three vertices. The bisectors of the angles $\angle BPC$, $\angle CPA$ and $\angle APB$ meet the sides BC, CA, and AB of triangle ABC at points D, E, and F, respectively. Unexpectedly, we find that AD, BE, and CF meet at a common point R. It should also be noted that the point P has another interesting property as it relates to the sides of triangle ABC, namely, $\frac{1}{2}(AB + BC + CA) < PA + PB + PC < 2(AB + BC + CA)$.

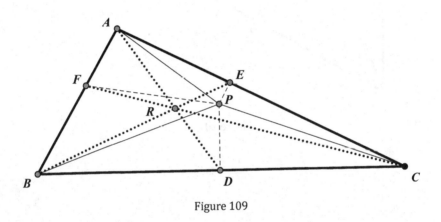

Figure 109

Curiosity 110. A Counterintuitive Concurrency

There are times when a concurrency is totally counterintuitive, as is the case with the configuration shown in Figure 110. Here, a random line is drawn external to triangle ABC. From the vertices of triangle ABC, perpendiculars AD, BE, and CF are drawn to this line, intersecting it at points D, E, and F. Now, here is the part that is quite astonishing, bearing in mind that the external line was randomly drawn. When perpendiculars are drawn from each of these three points on the external line to the corresponding sides of the original triangle ABC, we find that they are concurrent at a point P. In other words, with $DG \perp BC$, $EH \perp CA$, and $FJ \perp AB$, lines DG, EH, and FJ are concurrent at point P.

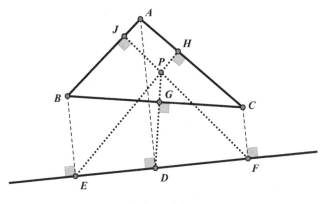

Figure 110

Curiosity 111. Concurrent Angle Bisectors of Triangles with a Common Base

In Figure 111 we show four triangles with a common base *AB* and equal vertex angles at points *C*, *D*, *E*, and *F*, namely, $\angle ACB = \angle ADB =$

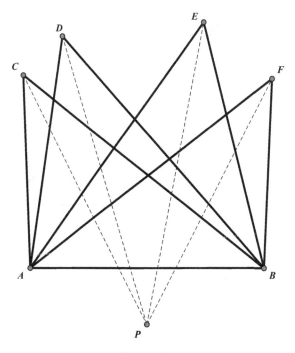

Figure 111

$\angle AEB = \angle AFB$. Quite unexpectedly, we find that the bisectors of each of the equal angles are concurrent at a point P.

Curiosity 112. An Unexpected Concurrency with the Circumscribed Circle

Strange and unexpected concurrencies can also occur with a circle circumscribed about a triangle. In Figure 112, triangle ABC is inscribed in the circle with center O. The bisector of $\angle BAC$ intersects side BC at point D. The line DP, which is perpendicular to BC at point D, intersects side AC at point J. Furthermore, the perpendicular bisector of AD intersects side AB at point K, intersects side AC at point L, and intersects line DJ at point P. Unforeseeably, the radius AO of the circumscribed circle also contains point P. We therefore have an unexpected concurrency of lines KL, DJ, and AO, which intersect at point P.

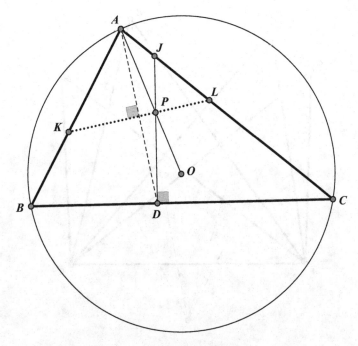

Figure 112

Curiosity 113. Another Surprising Aspect of Concurrent Cevians

Randomly drawn concurrent Cevians (i.e., line segments joining triangle vertices to points on their opposite sides) present a rather surprising result when we construct parallel lines to each of the other two Cevians through their respective feet. In Figure 113, we begin with the three Cevians, *AD*, *BE*, and *CF*, which intersect at point *U*. We then draw the following parallels: *AD∥FQ∥ER*, *BE∥DY∥FX*, and *CF∥EV∥DW*. The intersections of these lines are then defined as follows: *FX* intersects *VE* at point *M*, *FQ* intersects *DW* at point *N*, and *ER* intersects *DY* at point *S*. We then find that the three lines *MD*, *NE*, and *SF* are concurrent at a point *P*, and bisect each other.

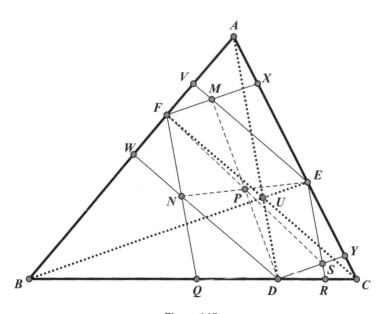

Figure 113

Curiosity 114. A Surprising Equality

Geometry tends to show unexpected relationships. One such is shown in Figure 114, where angle A of triangle ABC is bisected by the line AD with D on the side BC. At point B, a line is drawn parallel to AD, intersecting CA extended at point E. The circle containing the points C, D, and E is intersected by AD extended at point G, while the circumscribed circle of triangle ABC is intersected by AD extended at point H. Completely by surprise, we find that $AG = AH$.

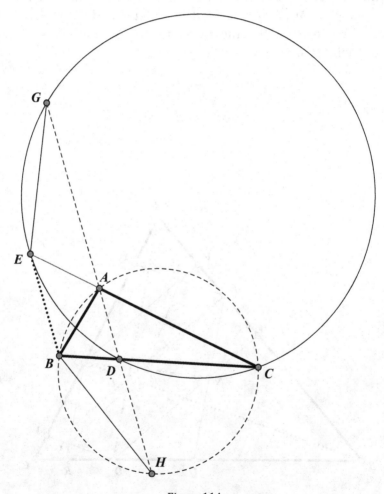

Figure 114

Curiosity 115. An Unforeseen Triangle Surprise

Another unforeseen triangle surprise occurs in triangle *ABC* as shown in Figure 115. Here, *M* is the midpoint of *BC* and *P* is a random point on *BC*. Perpendiculars *BK* and *CH* are drawn from points *B* and *C* to line *AP* (extended). Unexpectedly, we find that *MK* = *MH*.

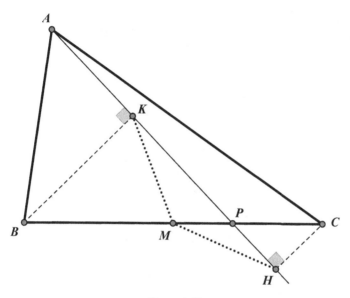

Figure 115

Curiosity 116. A Remarkable Property of Two Triangles with a Common Base

Two triangles that share the same base and whose vertices lie on a line parallel to the base have a rather interesting characteristic. Any line parallel to the base will have equal segments cut by the remaining two sides of each of the triangles. In Figure 116(a), $\triangle ABC$ and $\triangle DBC$ share the same base *BC* and vertices *A* and *D* lie on a line parallel to *BC*. Drawing a randomly selected line *EH* parallel to *BC*, we obtain *EF* = *GH*. This is true for any such parallel line and choosing another randomly-selected line *IL* parallel to *BC*, we once again find that the two triangles partition equal segments on the line *IL*, with *IK* = *JL* (and *IJ* = *LK*).

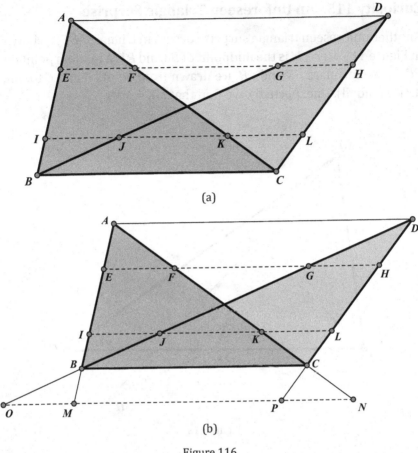

(a)

(b)

Figure 116

In fact, this wondrous partition of parallel lines can also be extended beyond that shown in Figure 116(a). In Figure 116(b), we extend the sides of the two triangles $\triangle ABC$ and $\triangle DBC$ to meet line ON, which is parallel to BC, at points M, N, O, and P. Once again, we find that $MN = OP$ (and $MO = PN$).

Curiosity 117. A Surprising Line Partitioning

The equality of line segments sometimes evolves when least expected and can be quite surprising. An example of this is shown in Figure 117.

Here, we are given a right triangle *ABC* with its right angle at point *C*. The midpoints *K*, *M*, and *N* of sides *BC*, *AC*, and *AB*, respectively, are the centers of circles with the sides of the right triangle as diameters. We then draw any line from point *C*, cutting each of the three circles once at a point other than *C*. Naming these intersections points *E*, *F*, and *G*, we discover a most astonishing equality along the line *CG*. Surprisingly, we have *CE = FG*.

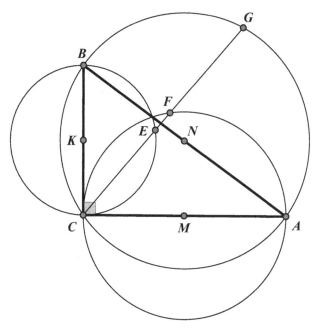

Figure 117

Curiosity 118. The Hidden Length Equality of Line Segments in a Triangle

To find line segments, the sum of whose lengths is the same as the length of a given line segment can be a surprising wonder. Consider the line segment *AH* in triangle *ABC* shown in Figure 118. From point *H* on *BC* we construct a parallel line to side *AB* intersecting side *AC* at point *E*. Similarly, we construct a line through point *H* parallel to side *AC* intersecting side *AB* at point *D*. We then draw lines *DF* parallel to

AH intersecting *BC* at point *F*, and *EG* parallel to *AH*, intersecting side *BC* at point *G*. Quite surprisingly, we find that *DF* + *EG* = *AH*.

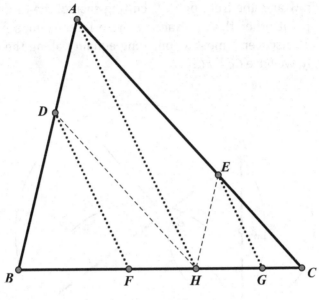

Figure 118

Curiosity 119. The Unexpected Angle Measure

Right triangles tend to hold many hidden surprises. Consider triangle *ABC*, shown in Figure 119, with altitude *CD*. The line *CF* is the bisector

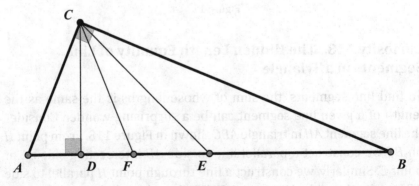

Figure 119

of angle ∠*ACB*, and *CE* is the median to side *AB*. Furthermore, we assume that ∠*ACD* = ∠*DCF* = ∠*FCE* = ∠*ECB*. Such being the case, the angle *ACB* must be a right angle! This is quite an unusual situation.

Curiosity 120. An Unexpected Equality from a Right Triangle

Right triangles can produce some very startling results, as we see in Figure 120, where the altitude drawn to the hypotenuse *AB* of right triangle *ABC* is extended through point *C* so that *CP* = *AC*. Hypotenuse *AB* is extended through point *A* to point *Q* so that *AQ* = *BC*. Quite unexpectedly, we find that *PB* = *QC*.

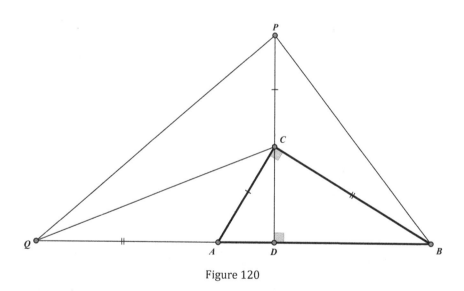

Figure 120

Curiosity 121. The Conundrum: Perpendicular or Parallel

There are times when the information provided for a geometric situation is not precise enough and can lead to various conclusions. One such example can be seen in Figures 121(a) and 121(b). In both figures, we begin with a right triangle *ABC* with hypotenuse *AB* and construct a point *P* such that *BP* ⊥ *AB* and *BP* = *BC*. We wish to compare the relative positions of *CP* and the angle bisector *AD* of ∠*BAC*. In Figure 121(a), the perpendicular *BP* is drawn to the right of the triangle, and when *PC* intersects the bisector *AD* of ∠*BAC*, it is perpendicular to it at point *E*. On the other hand, in Figure 121(b), with the perpendicular *BP* drawn in the other direction, yet also perpendicular to *AB* and equal to *BC*, we see that *PC* is parallel to the bisector *AD* of ∠*BAC*.

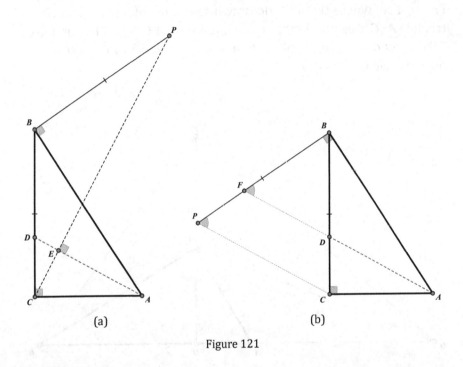

(a)   (b)

Figure 121

Curiosity 122. Multiple Midpoints in a Right Triangle Determine Equal Line Segments

We present a rather unusual development of equality through multiple midpoints inside a right triangle. In Figure 122, points *D* and *E* are any randomly selected points on sides *AB* and *AC*, respectively, of triangle *ABC*. We then consider the midpoints *F*, *G*, *H*, and *J* of segments *DE*, *BE*, *DC*, and *BC*, respectively. Most unexpectedly, we find that *FJ* = *GH*.

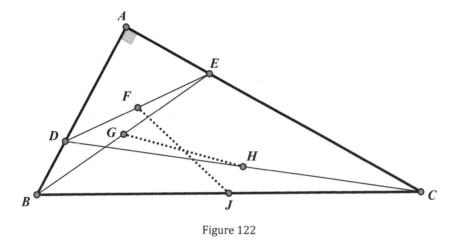

Figure 122

Curiosity 123. Perpendiculars in Right Triangles that Generate Equal Angles

There are further surprises that right triangles seem to harbor, such as the one we see in Figure 123, where right triangle *ABC* with altitude *CD* on the hypotenuse *AB* has perpendicular lines drawn from point *D* to the right triangle's legs *CA* and *CB* at points *E* and *F*, respectively. Completely unexpectedly, we find that ∠*AEB* = ∠*AFB*.

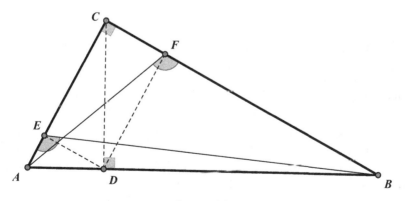

Figure 123

Curiosity 124. Right Triangles Sharing a Common Hypotenuse Generate an Unexpected Equality

A conglomeration of perpendicular lines is another way to produce some unexpected results. Consider two non-congruent right triangles sharing the same hypotenuse, as we show in Figure 124, where right triangle *ACB* and right triangle *ABD* share the hypotenuse *AB*, with midpoint *M*. When we construct perpendiculars to *MC* and to *MD*, they meet at a point *P*. We now find that *PC* = *PD*. This amazing relationship certainly seems to appear unexpectedly.

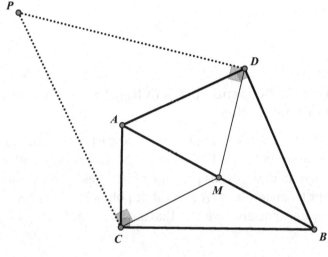

Figure 124

Curiosity 125. The Pythagorean Theorem Revisited Geometrically

Placing squares on the sides of a right triangle geometrically explains the famous Pythagorean theorem: the sum of the squares of the lengths of the legs of a right triangle is equal to the square of the length of the hypotenuse. The property is commonly written as $a^2 + b^2 = c^2$. Geometrically, this could be interpreted in the following way: the sum of the areas of the squares on the legs of a right triangle is equal to the area of the square on the hypotenuse, as we see in Figure 125.

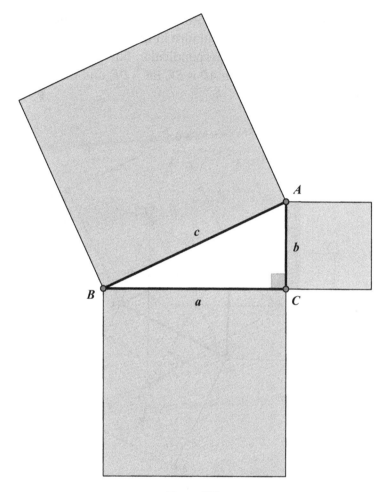

Figure 125

Curiosity 126. Squares on Triangle Sides Produce Noteworthy Surprises

We need not limit ourselves to right triangles. Placing squares on each of the sides of a randomly drawn triangle *ABC*, as we show in Figure 126, also produces some interesting surprises. We obtain concurrent lines by joining the center of each square with the remote vertex of the original triangle *ABC*. Furthermore, we find that these three lines, *AD*, *BE*, and *CF*, are perpendicular to the lines joining the midpoints of adjacent squares. In other words, $AD \perp EF$, $BE \perp DF$, and $CF \perp DE$. There is even more to admire in this configuration, namely, that these lines are not only perpendicular, but also equal to each other. In other words, we have $AD = EF$, $BE = DF$, and $CF = DE$. Truly unexpected occurrences!

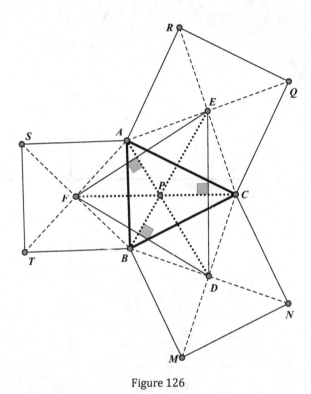

Figure 126

Curiosity 127. More Properties Generated by Squares on Triangle Sides

The squares on the sides of a randomly selected triangle produce many other extraordinary surprises. Let's once again consider triangle ABC with the squares on each side as we can see in Figure 127. This time we recognize two parallelograms, $BTXM$ and $CNYQ$. The astonishing result is that we have created an isosceles right triangle AXY where $\angle XAY = 90°$, $AX = AY$, and $AX \perp AY$.

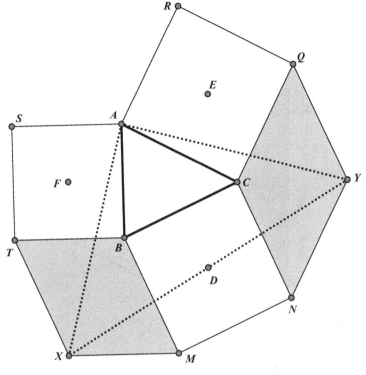

Figure 127

Curiosity 128. An Unexpected Perpendicularity

There is also another perpendicularity in the configuration of Figure 127. As we see in Figure 128, we find that *CT* is perpendicular to *AM*. There is yet more to be found in this geometric arrangement, as we will see in the following curiosities.

Curiosity 129. A Surprising Concurrency from Squares on the Sides of a Triangle

We now draw the third parallelogram *ARZS* and then join midpoints of the parallelograms with the remote squares' centers, as shown in Figure 129. Unexpectedly, this yields three concurrent lines, namely, *DG*, *EJ*, and *FH*.

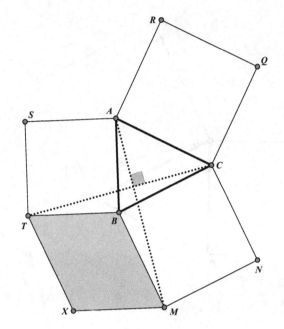

Figure 128

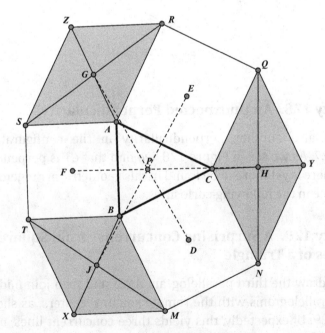

Figure 129

Curiosity 130. Another Surprising Concurrency from Squares on Triangle Sides

This configuration yields further concurrencies. If we consider the lines *GU*, *JV*, and *HW* joining the center point of each parallelogram (namely, *G*, *H*, and *J*) with the midpoints of the remote sides of the original triangle *ABC*, we find that these lines are concurrent at the point *L*, as we see in Figure 130.

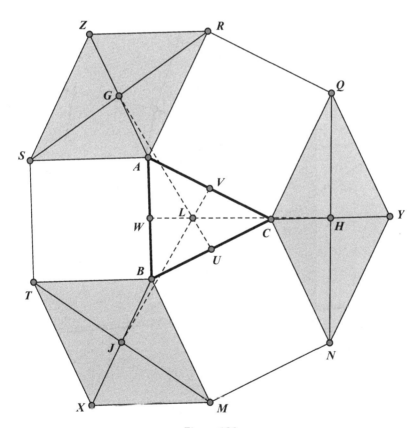

Figure 130

Curiosity 131. Using Squares on Triangle Sides to Create another Surprising Concurrency

Placing the squares on the sides of a triangle continues to lead to a variety of other concurrencies. This time, we extend the external sides of the squares placed on triangle *ABC* until they meet at points *X*, *Y*, and *Z*. This creates the triangle *XYZ*, shown in Figure 131. By joining each of these points to the nearest vertices of triangle *ABC*, we find once again that a concurrency emerges at point *P*.

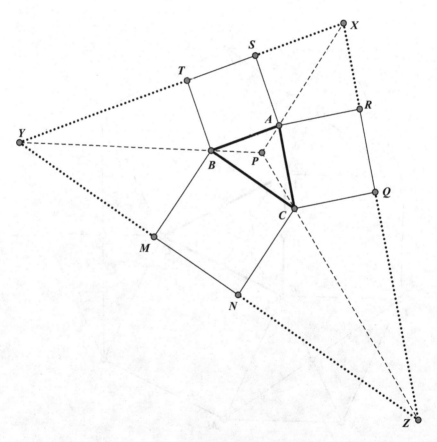

Figure 131

Curiosity 132. Squares on the Legs of a Right Triangle Produce an Unexpected Equality

When we place a square on each of the legs of a right triangle and add some simple connections, we find some unexpected surprises. We see this in Figure 132, where right triangle *ABC* has square *ACDE* and square *CBFG* on sides *CA* and *BC*, respectively. We draw *AF* and *BE* intersecting in *P*, with *BE* and *CA* intersecting in *K* and *AF* and *BC* intersecting in *J*. Quite unexpectedly, we find that *CK* = *CJ*, regardless of the shape of the right triangle.

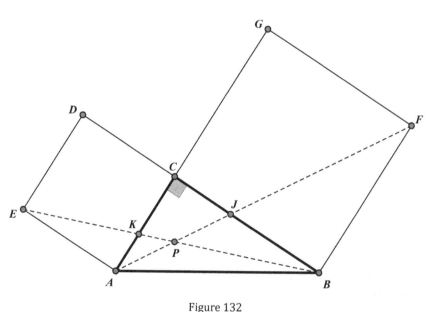

Figure 132

Curiosity 133. More About Squares on the Legs of a Right Triangle

Some more unusual results can be found in this configuration, as shown in Figure 133. Once again, we have the squares *ACDE* and *CBFG* placed on sides *CA* and *BC*, respectively. From vertex *E* the perpendicular is drawn to the extension of *AB* intersecting it at point *S*, and from vertex *F* a perpendicular is drawn to the extension of *AB* intersecting it at point *T*. A first unexpected result is *ES* + *FT* = *AB*. Furthermore, we also see that area[*AES*] + area[*FBT*] = area[*ABC*].

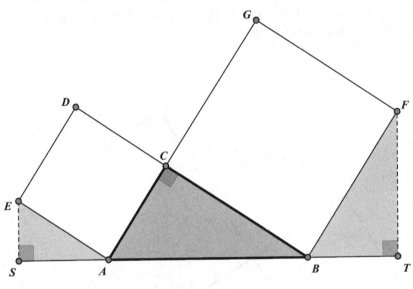

Figure 133

Curiosity 134. More Placements of Squares on Right Triangles

Placing a square on a leg and a square on the hypotenuse of a right triangle produces some more unexpected results, as we see in Figure 134. Here, right triangle *ABC* has square *BFGC* placed on side *BC* and square *AEDB* placed on hypotenuse *AB*. We select a point *L* anywhere on *BC*. Quite counterintuitively, we find that area[*LBF*] = area[*LDB*]. What adds to the remarkability of this result is that the equality of the areas of these two triangles remains constant, regardless of the location of point *L* on line *BC*.

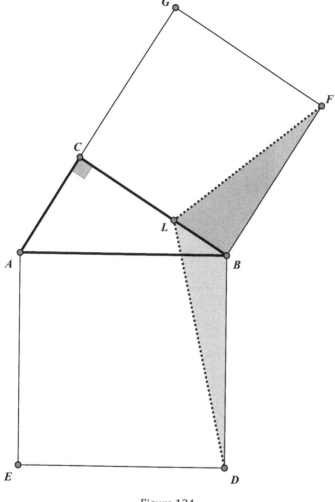

Figure 134

Curiosity 135. Another Unexpected Area Equality

A similar configuration to the one shown in Figure 134 yields a different pair of triangles that provide another similarly astonishing result. In Figure 135, we have once again placed a square *AEDB* on the hypotenuse of triangle *ABC*. Somewhat surprisingly, we find that area$[ACE]$ + area$[BCD]$ = $\frac{1}{2}$area$[AEBD]$.

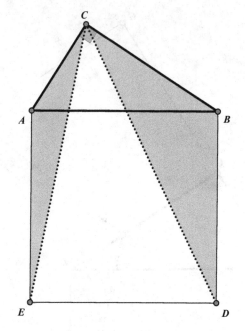

Figure 135

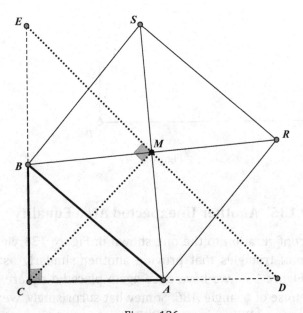

Figure 136

Curiosity 136. The Square on the Hypotenuse of a Right Triangle

In Figure 136, we once again place a square *ARSB* on the hypotenuse *AB* of right triangle *ABC*. We then draw a line joining its midpoint *M* with the vertex *C*. The line through *M* perpendicular to *CM* intersects the extensions of the two sides *CA* and *CB* at points *D* and *E*, respectively. The unexpected result is that *BE* = *AC* and *AD* = *BC*. Furthermore, another surprising feature of this configuration is that *CM* bisects angle ∠*ACB*.

Curiosity 137. A Truly Unexpected Collinearity

Placing an equilateral triangle inside a given square *ABCD* and another equilateral triangle outside the square, as shown in Figure 137, provides an interesting and unexpected collinearity of three points. The equilateral triangle *BCE* and equilateral triangle *DCF* are placed inside and outside of square *ABCD*. The result is that the points *A*, *E*, and *F* are collinear.

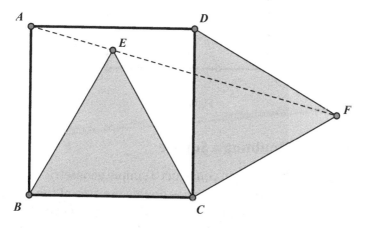

Figure 137

Curiosity 138. A Most Unusual Procedure to Divide a Square into Two Equal Parts

It is rather easy to divide a square into two equal parts. Two simple ways are by drawing the diagonal of the square, or by joining the midpoints of a pair of opposite sides of the square. The procedure we show in Figure 138 is rather unusual as it demonstrates the beauty of mathematics when it is least expected. We begin by drawing any right triangle *AED* with its hypotenuse on the side *AD* of square *ABCD*. The bisector of angle *AED*, which intersects sides *AD* and *BC* at points *K* and *L*, respectively, partitions the square into two equal area regions. In other words, the area of quadrilateral *ABLK* is equal to the area of quadrilateral *DKLC*.

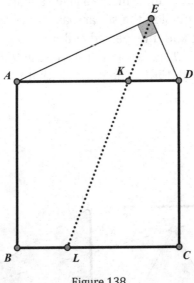

Figure 138

Curiosity 139. Doubling a Square

It is astounding when we can construct a simple geometric figure that results in two squares, where one has double the area of the other. We begin with triangle *ABC* with ∠*A* = 45°, shown in Figure 139, where perpendiculars from vertices *B* and *C* are drawn to the opposite sides intersecting them at points *D* and *E*, respectively. Completely unexpectedly, we find that the square on *BC*, namely, *BGFC*, has exactly twice the area of the square *DHIE*.

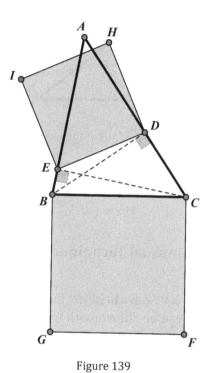

Figure 139

Curiosity 140. A Strange Construction of Parallel Lines

There are curious ways to construct a line parallel to the side of a triangle. One way is to first construct the median to one side of the triangle and then draw lines from the two endpoints of that side in such a way that they intersect each other on the median. This is shown in Figure 140, where *AD* is a median of triangle *ABC*. Suppose we would like to construct a line through point *P* parallel to *BC*. All we would need to do is draw line *PC*, which intersects the median *AD* at point *G*. Then, by drawing a line from *B* through *G* to meet *AC* at *Q*, we will construct *PQ* parallel to *BC*.

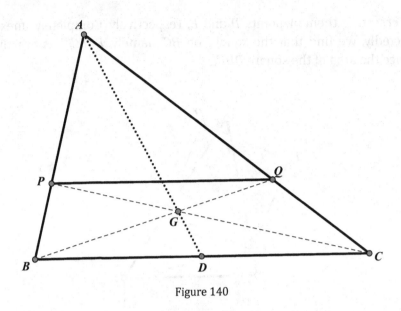

Figure 140

Curiosity 141. An Unusual Technique to Find the Midpoint of a Line Segment

There are also very curious ways to bisect lines. A technique related to the previous example can be shown with triangle *ABC* in Figure 141.

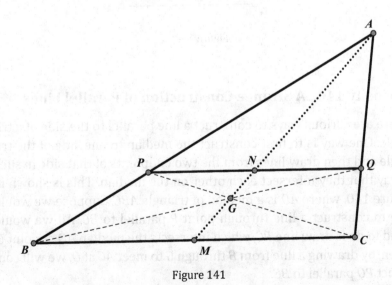

Figure 141

Here, we are given a line segment *BC* that we wish to bisect. We choose a point *A* anywhere not on the extension of *BC*. We then draw a line *PQ* parallel to *BC* with point *P* on *AB* and point *Q* on *AC*. We then draw lines *PC* and *QB*, which intersect at a point *G*. When we draw *AG* extended, we find that this line bisects both *PQ* and *BC,* at points *N* and *M*, respectively, so that *PN* = *NQ*, and *BM* = *MC*.

Curiosity 142. The Unexpected Appearance of Parallel Lines

Parallel lines sometimes appear quite unexpectedly, once again demonstrating interesting aspects of geometry. An example of this, shown in Figure 142, results by drawing any line containing vertex *A* of triangle *ABC*, whose sides *BC*, *AC*, and *AB* have midpoints *D*, *E*, and *F*, respectively. The line *DE* extended meets that randomly-selected line through vertex *A* at point *G*, and the line *DF* extended meets that same line at point *H*. Surprisingly, we find that *GC* is parallel to *HB*, once again exposing a surprising hidden symmetry.

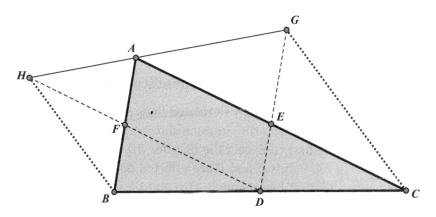

Figure 142

Curiosity 143. The Unanticipated Parallel Line

Another surprising result is the following. In Figure 143, we see triangle *ABC* and the angle bisectors *BL* and *CK* of the angles ∠*CBA* and ∠*ACB*, respectively. When we construct the perpendiculars from point *A* to each of the bisectors (extended if necessary), we have *AD* ⊥ *BL* at point *D* and *AE* ⊥ *CK* at point *E*. Completely unexpectedly, we find that the line segment *DE* is parallel to the base *BC* of triangle *ABC*.

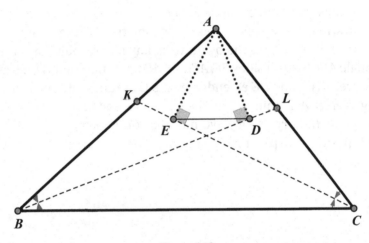

Figure 143

Curiosity 144. The Surprising Perpendicularity

There are times when seemingly unrelated lines prove to be perpendicular. That is the case with the situation shown in Figure 144, where in triangle *ABC* we first construct the bisector *AD* of angle *BAC*, with point *D* on side *BC*. Points *E* and *F* are selected on sides *AB* and *AC*, respectively, so that *DE* bisects angle *FDB* and *DF* bisects angle *CDA*. When *EF* is extended, it intersects *BC* extended at point *G*. Completely by surprise, we find that *GA* is perpendicular to *AD*.

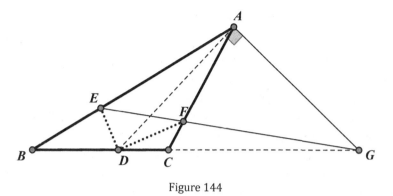

Figure 144

Curiosity 145. Another Unexpected Perpendicularity

In Figure 145, the altitude *AH* in triangle *ABC* intersects the side *BC* at point *H*. Also, *M* is the midpoint of side *BC*. A point *Q* is determined, so that ∠*AQC* = 90°, and a point *P* on *HQ* is then determined so that ∠*APB* = 90°. When the point *R* is selected as the midpoint of *PQ*, the angle *ARM*, quite unexpectedly, also turns out to be a right angle.

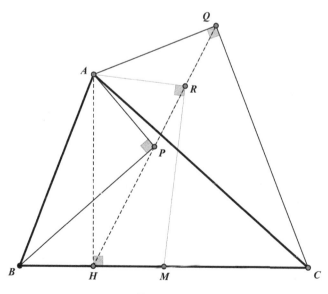

Figure 145

Curiosity 146. Yet Another Unexpected Right Angle

As in the previous example, there are some surprising procedures that produce right angles when least expected. In Figure 146, we begin with triangle *ABC* and a randomly selected point *P* on side *BC*. We first locate the midpoint *O* of side *BC* and construct a circle with center *O* and radius *OC*. This circle intersects *AP* extended at point *F*. Next, we determine the line through point *P* that is parallel to *FO*. This line *PN* intersects *AO* at point *N*. Then, we draw the line parallel to *BC* through point *N*, intersecting sides *AB* and *AC* at points *J* and *H*, respectively. We are then astonished to find that ∠*HPJ* is quite unexpectedly also a right angle.

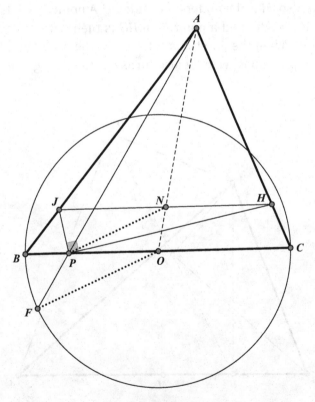

Figure 146

Curiosity 147. Four Important Concyclic Points

Finding four or more points that all lie on the same circle is always a noteworthy consideration in geometry. The extension of an altitude of a triangle can have a surprising result. Suppose we are given a triangle *ABC*, as we can see in Figure 147, with *H* as its orthocenter and *P* as the intersection of altitude *AH* with side *BC*. When we extend *HP* its own length to point *K*, amazingly, we find that this point *K* also lies on the circumscribed circle of triangle *ABC*. This produces a circle with four points on it, namely, *A*, *B*, *K*, and *C*.

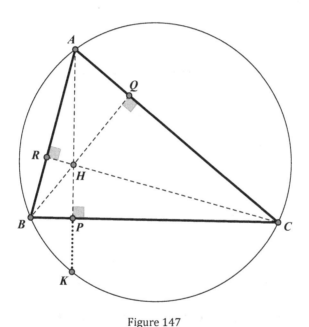

Figure 147

Curiosity 148. Four Remarkable Concyclic Points

Concyclic points tend to appear when least expected. Consider triangle *ABC* in Figure 148 with its circumscribed circle's center at point *O*. The points *E* and *F* are placed on sides *AB* and *AC* so that *EF* will be perpendicular to *AO*. Once that is set, we find that the points *B*, *E*, *F*, and *C* are concyclic.

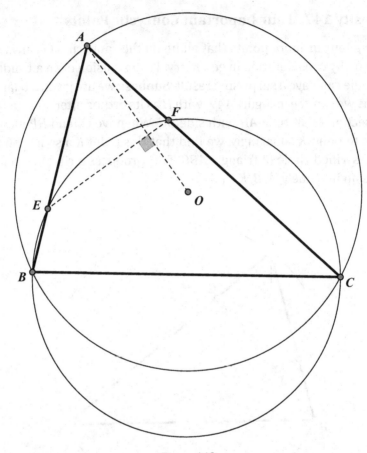

Figure 148

Curiosity 149. Four Unexpected Concyclic Points

As we have been saying, finding four or more points that lie on the same circle is always a challenge in geometry. Another such example is shown in Figure 149, where H is the orthocenter of triangle ABC and points F and D are the midpoints of sides AB and BC, respectively. When we extend DH to intersect the circumscribed circle of triangle ABC, the point of intersection is point X. Extending FH, it intersects the circumscribed circle at point Y. Astonishingly, the points D, F, X, and Y lie on the same circle.

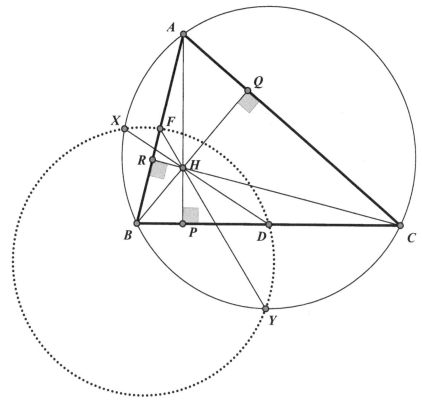

Figure 149

Curiosity 150. Perpendiculars that Generate Concyclic Points

Yet another collection of perpendicular lines in a triangle will generate concyclic points. In Figure 150, right triangle *ABC*, with its right angle at point *A*, has two randomly selected points *P* and *Q* on sides *AB* and *AC*, respectively. We now construct perpendicular lines from point *A* to lines *BC*, *PC*, *BQ*, and *PQ*, intersecting these lines at points *E*, *G*, *H*, and *J*, respectively. Surprisingly, we find that these points *E*, *G*, *H*, and *J* all lie on the same circle.

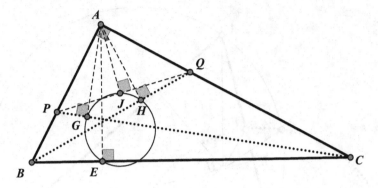

Figure 150

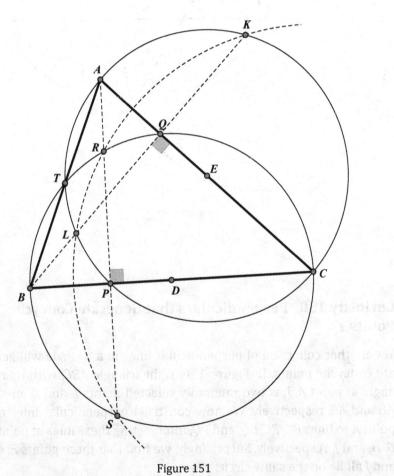

Figure 151

Curiosity 151. Altitudes and Circles that Generate Another Circle

Intersecting circles can also produce other concyclic points. In Figure 151, sides *BC* and *AC* of triangle *ABC* are the diameters for circles with centers at points *D* and *E*, respectively. Altitude *AP* extended intersects the circle with center *D* at points *R* and *S*. Altitude *BQ* intersects the circle with center *E* at points *L* and *K*. These four intersection points, namely, *S*, *L*, *R*, and *K*, all lie on the same circle, only a portion of which is shown in Figure 151. In this configuration, it should also be noted that the two circles with centers at *D* and *E* intersect at a point *T* that lies on side *AB*.

Curiosity 152. More Unexpected Concyclic Points

This time, several perpendiculars in an acute triangle will generate four concyclic points. We begin Figure 152 by selecting a point *R* in the interior of triangle *ABC*. From this point, we draw perpendiculars to each of the three sides, intersecting *AB*, *BC*, and *AC* at points *F*, *D*, and

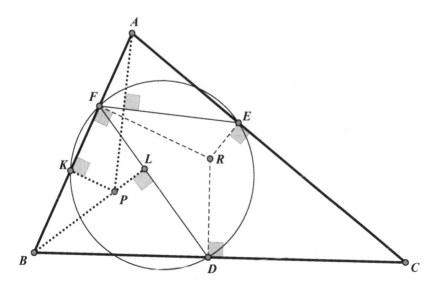

Figure 152

E, respectively. We then construct the perpendicular from point *B* to line *DF*, intersecting it at point *L*. The perpendicular from point *A* to line *EF* when extended meets line *BL* at point *P* and the perpendicular from point *P* to line *AB* intersects *AB* at point *K*. Unexpectedly, all of these perpendiculars yield four concyclic points, namely, points *D*, *E*, *F*, and *K*.

Curiosity 153. A Surprising Five-Point Circle

Having five points on the same circle is quite astonishing. We have that in Figure 153, where point *H* is the orthocenter of △*ABC*, with altitudes *AP*, *BQ*, and *CR*. The points *E*, *F*, and *M* are the midpoints of *BQ*, *CR*, and *BC*, respectively. Surprisingly, the points *E*, *H*, *F*, *D*, and *M* all lie on the same circle, providing a 5-point circle. As an extra marvel, we find that △*PFE* is similar to △*ABC*.

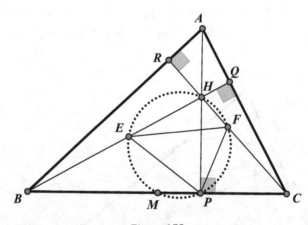

Figure 153

Curiosity 154. The Famous Nine-Point Circle

It is certainly to be admired when four specific points lie on the same circle. Sometimes, however, this can occur for far more than four points. One of the most famous relationships in plane geometry is the Nine-Point Circle, which is, as the name indicates, a circle that has nine specific triangle-related points on it. Initially, in 1765, the famous Swiss mathematician Leonhard Euler (1707–1783) discovered that there are six triangle-related points that lie on a circle. These points are the midpoints of the sides of a triangle, which in Figure 154 are the points *K*, *M*, and *N*, as well as the points that are the feet of the three altitudes, namely, points *D*, *E*, and *F*. This was already quite an amazing discovery – but there was more to come!

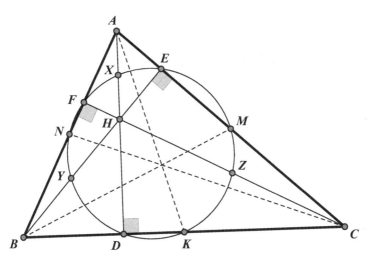

Figure 154

In 1822, the German mathematician Karl Wilhelm Feuerbach (1800–1834) discovered that another three points also lie on this same Euler circle, thus making it a Nine-Point Circle. In Figure 154, these additional points, *X*, *Y*, and *Z*, are the midpoints of the segments *AH*, *BH*, and *CH*, which each join a vertex with the orthocenter *H* of the triangle *ABC*.

In summary, we see that the Nine-Point Circle contains nine specific points of the triangle *ABC*, namely, the feet of the altitudes, *D*, *E*, and *F*, the midpoints of the sides, *K*, *M*, and *N*, and the midpoints of the segments joining each vertex with the orthocenter, *X*, *Y*, and *Z*.

Curiosity 155. A Collinearity with the Center of the Nine-Point Circle

An interesting collinearity is shown in Figure 155 for triangle *ABC*. The foot *S* of the perpendicular from the orthocenter *H* to the bisector *AT* of ∠*BAC*, the center *R* of the Nine-Point Circle, and the midpoint *K* of the opposite side *BC* all lie on the same line. In other words, the points *S*, *R*, and *K* are collinear.

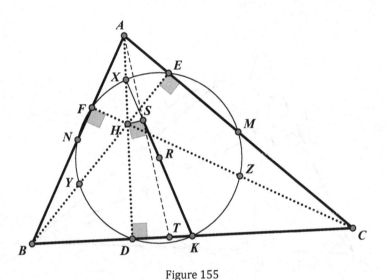

Figure 155

Curiosity 156. The Meeting of the Three Famous Triangle Centers

There is a special significance about some of the points in a triangle, namely, the orthocenter (the point of intersection of the altitudes of the triangle), the center of the circumscribed circle (the point of intersection of the perpendicular bisectors of the the sides of the triangle), and the centroid (the point of intersection of the medians of a triangle). In Figure 156, these are designated by *H*, *O*, and *G*, respectively. Not only are these three points always collinear, but also the following relationship always holds true: $OG = \frac{1}{2}HG$.

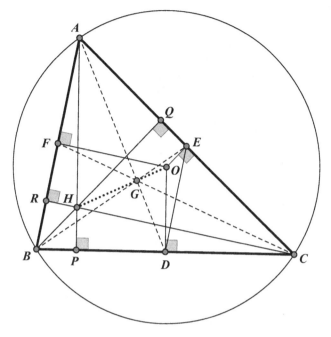

Figure 156

Curiosity 157. Properties of the Nine-Point Circle

In Curiosity 156, we have already established that points H, O, and G are collinear with $OG = \frac{1}{2}HG$. In Figure 157, we see that the center of the Nine-Point Circle, R, is the midpoint of the line segment HO, and we have, thus, established another collinearity of four points, namely, H, R, G, and O. Furthermore, the radius of the Nine-Point Circle is half the length of the radius of the circumscribed circle.

Curiosity 158. More Properties of the Nine-Point Circle

There is yet another spectacular feature that the Nine-Point Circle provides for us. Any line segment from the orthocenter to the circumscribed circle is bisected by the Nine-Point Circle. Suppose we take a random line segment HJ in Figure 158 which joins the orthocenter H of triangle ABC with a point J on the circumscribed circle. The point L

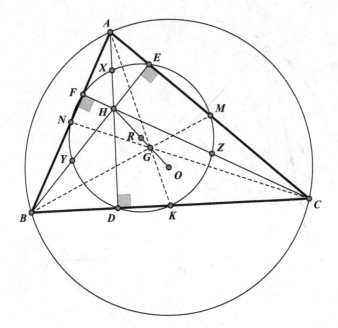

Figure 157

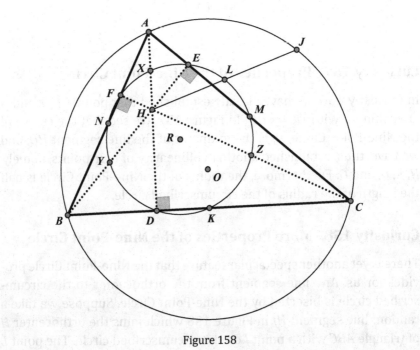

Figure 158

is where it intersects the Nine-Point Circle. We find that *HJ* is bisected at point *L*, so that *HL* = *LJ*.

Curiosity 159. An Unexpected Collinearity

The feet of the altitudes of a triangle can sometimes lead to a most unexpected result, as we will see in Figure 159, where a circle is drawn containing vertex *A* and the feet of altitudes *BE* and *CF* of triangle *ABC*, namely, points *E* and *F*. The circle through points *A*, *E*, and *F* must also contain the intersection, point *H*, of lines *BE* and *CF*. Since quadrilateral *AFHE* has a pair of supplementary opposite angles as ∠*AEH* = ∠*HFA* = 90°, we have quadrilateral *AFHE* as a cyclic quadrilateral. However, the amazing surprise is that the tangents to the circle at points *E* and *F* will intersect at point *X*, which is on side *BC*, regardless of the shape of the original triangle. We can then also say that the points *B*, *X*, and *C* are collinear.

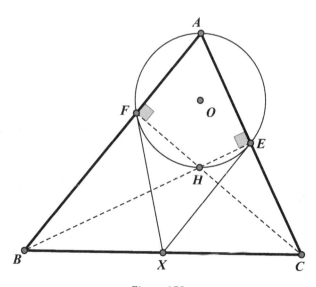

Figure 159

Curiosity 160. A Concurrency Generated by the Orthic Triangle

In a rather unexpected way, when the orthic triangle gets involved with the midpoints of the opposite sides of a triangle, perpendiculars appear. Consider triangle *ABC* in Figure 160 with its orthic triangle *DEF* (which is the triangle formed by the feet of the altitudes of the triangle). Points *M*, *N*, and *R* are the midpoints of the sides of triangle *ABC*. When we draw the perpendiculars from each of these midpoints to the opposite sides of the orthic triangle, lo and behold, these three lines are concurrent at a point *P*.

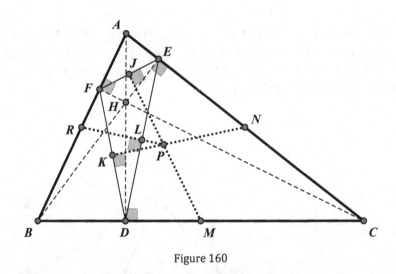

Figure 160

Curiosity 161. Altitudes Produce a Concurrency and Equality

In Figure 161, the altitudes *AE*, *BF*, and *CD* intersect at point *H*, the orthocenter of triangle *ABC*. We locate the midpoints *T*, *U*, and *V* of line segments *AH*, *BH*, and *CH*, respectively. We then locate the midpoints *K*, *M*, and *N*, of the sides *BC*, *CA*, and *AB* of triangle *ABC*, respectively. The result is that the line segments *TK*, *UM*, and *VN* are concurrent in a point *P*. Additionally, these line segments are also equal in length as *TK = UM = VN*. Recall the Nine-Point Circle!

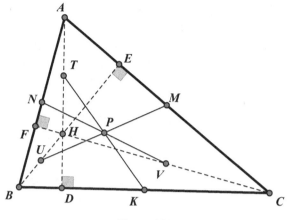

Figure 161

Curiosity 162. Napoleon's Contribution to Mathematics

The equilateral triangle can also play an important role in conjunction with a randomly drawn triangle. Suppose we draw equilateral triangles *ADB*, *BEC*, and *CFA* on the sides of a randomly-drawn triangle *ABC*,

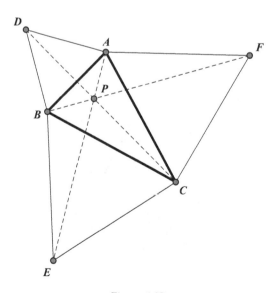

Figure 162

as we do in Figure 162. When we draw lines *AE*, *BF*, and *CD*, we find that they are equal in length and also concurrent at a point *P*. This is quite an amazing feature, since we started off with any randomly drawn triangle *ABC*, which led us to an equality and a concurrence. This relationship has been historically attributed to the Emperor Napoleon Bonaparte (1769–1821), even though there is some doubt as to whether he himself developed it or whether this was the work of his artillery engineer, Jean-Victor Poncelet (1788–1867).

Curiosity 163. Napoleon's Minimum Distance Point (the Fermat Point)

As if this weren't enough, the point *P* has other interesting character-istics as well. The point *P* is the location within the triangle where the sum of the distances to the three vertices is a minimum. This allows us to refer to the point *P* as the *minimum distance point*, also known as the *Fermat Point*, of the triangle. Furthermore, the point *P* is also the equiangular point of the triangle, since $\angle APB = \angle BPC = \angle APC$ (see Figure 163).

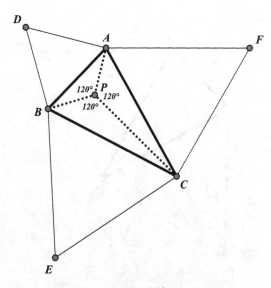

Figure 163

Curiosity 164. When the Minimum Distance Point is not Inside the Triangle

Suppose we draw a triangle *ABC*, in which one angle is equal to or exceeds 120°. If the angle is equal to 120°, we can see that point *P* coincides with the vertex *A* of the 120° angle, as shown in Figure 164(a). If this angle ∠*BAC* increases beyond 120°, as in Figure 164(b), the point *P* will be outside the triangle. Therefore, in order for our minimum distance point to be inside the triangle, we must ensure that the triangle has no angle greater than 120°.

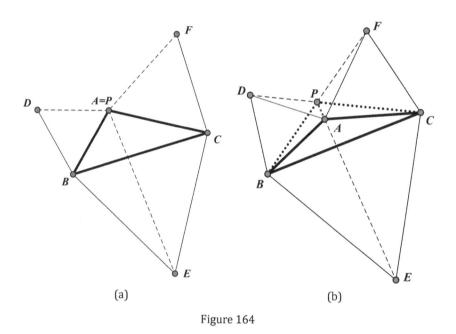

(a)   (b)

Figure 164

Curiosity 165. Extensions of Napoleon's Theorem

Another surprising feature of this configuration arises when we take the center of each of the three equilateral triangles that were drawn on the sides of triangle *ABC*, namely, points *K*, *L*, and *M*, and join them, as we have in Figure 165. The resulting triangle *KLM*, amazingly, turns out to be another equilateral triangle.

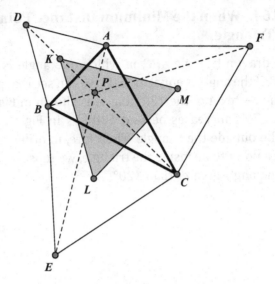

Figure 165

Curiosity 166. Overlapping Side-Equilateral Triangles

In case you're wondering whether we are required to construct the equilateral triangles *outside* the given triangle *ABC* to get these wonders, we show in Figure 166 what happens when the equilateral triangles on each of the sides of the original triangle *ABC* overlap this

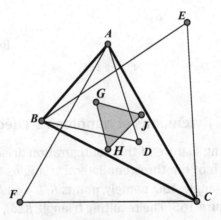

Figure 166

original triangle. To everyone's amazement, we still end up with an equilateral triangle *GHJ* by joining the midpoints of the three equilateral triangles, even though they are no longer external to the original triangle but overlap it.

Curiosity 167. Surprising Triangle Area Relationship

Combining these previous situations, a most unusual surprise emerges, as we can see in Figure 167. Here, we have drawn the three triangles from Figures 165 and 166, namely, $\triangle ABC$, $\triangle KLM$, and $\triangle GHJ$. The relationship between these three triangles is area[*KLM*] − area[*GHJ*] = area[*ABC*].

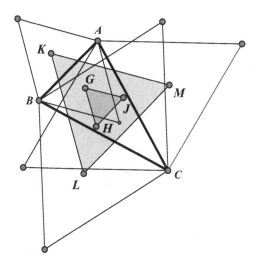

Figure 167

Curiosity 168. The Centroid Enters the Previous Configuration

Much to our surprise, there are more unexpected wonders in the configuration of equilateral triangles on the sides of a randomly-selected triangle. If we locate the centroid (the point of intersection of the medians) of triangle *ABC*, as we show in Figure 168, we find that it is exactly at the midpoint of the equilateral triangle *KLM*. If this were all, we would already have enough to keep us in awe. However, as we will see moving along, there are still more surprises left to be appreciated within this configuration.

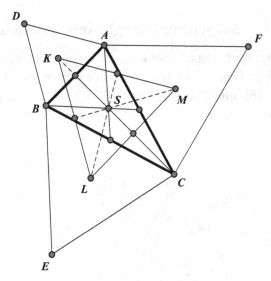

Figure 168

Curiosity 169. The Emergence of Another Equilateral Triangle

With a little creativity, we can construct another equilateral triangle in addition to the three equilateral triangles drawn on the sides of a given triangle *ABC*. As shown in Figure 169, we construct a parallelogram *ADCR* by drawing *AR* parallel to *DC* and *CR* parallel to *DA*. It then just so happens that △*AER* is also an equilateral triangle.

Curiosity 170. A Novel Way of Finding the Center of the Circumscribed Circle

All triangles have circumcircles. The usual way of locating the center of the circumscribed circle about a triangle *ABC* would be to construct the perpendicular bisectors of each of the three sides. Their common

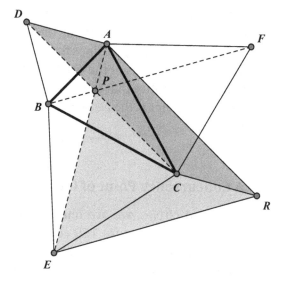

Figure 169

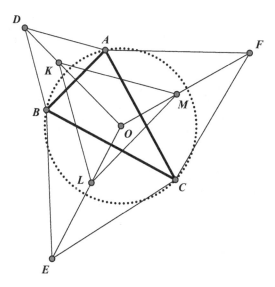

Figure 170

point is then the circumcenter. However, in Figure 170, we show an interesting alternative to this construction. Using the equilateral triangle *KLM* from Figure 165 and drawing the lines from the three vertices of this triangle to the remote vertices of the original three equilateral triangles (*KD*, *LE*, and *MF*), we notice that these three lines are concurrent and that their common point is the center *O* of the circumscribed circle of triangle *ABC*.

Curiosity 171. A Concurrency Point of Circles

Speaking of circumscribed circles, we are once again amazed that the minimum distance point *P* in Figure 169 comes into play in yet another way. We find that the circumscribed circles of each of the three original equilateral triangles drawn on the sides of triangle *ABC* are also concurrent at this very same point *P*, as we can see in Figure 171.

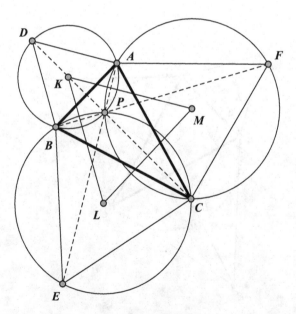

Figure 171

Curiosity 172. The Famous Miquel Theorem

Circles can also play a curious role in creating a concurrency. Suppose one were to take any three points *D*, *E*, and *F* on the sides of a randomly drawn triangle *ABC*, as shown in Figure 172, and then construct a circle using one vertex and a previously selected point on each of the adjacent sides. Doing this three times, we find that, regardless of where those previously selected three points were situated on the sides of the triangle, the three circles will always be concurrent at a point *P*. This point is named after the French mathematician Auguste Miquel (1816–1851), who published this relationship in 1838.

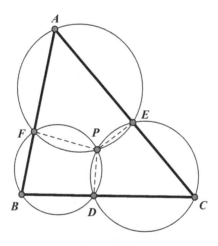

Figure 172

Curiosity 173. Miquel's Similar Triangles

As we see in Curiosity 172, the Miquel circles can take any position as long as they contain a vertex and a point on each of the two adjacent sides. A most astonishing result occurs when we connect the centers *Q*, *R*, and *S* of the three circles. This surprising result, shown in Figure 173, is that △*ABC* and △*QRS* are similar with the following angle equalities: ∠*A* = ∠*Q*, ∠*B* = ∠*R* and ∠*C* = ∠ *S*.

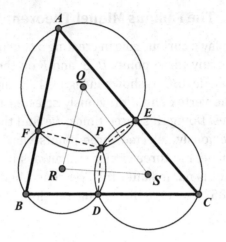

Figure 173

Curiosity 174. The Astounding Morley's Theorem

In 1900, the mathematician Frank Morley (1860–1937) discovered a most amazing relationship that applies to all triangles, regardless

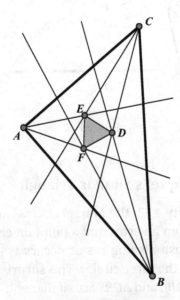

Figure 174

of their shape. Morley showed that if we trisect each of the angles of a triangle, the intersections of the adjacent trisectors determine an equilateral triangle, regardless of the shape of the original triangle. In Figure 174, we have each angle of triangle *ABC* trisected with the adjacent trisectors meeting at points *D*, *E*, and *F*. When we join these three points, the resulting triangle *DEF* is always equilateral.

Curiosity 175. Morley's Theorem Extended

As if Morley's theorem were not spectacular enough, there is still another amazing aspect of this configuration. When we draw the line segments joining the vertices of the newly formed equilateral triangle *EFD* with the remote vertices of the original triangle *ABC*, the lines are concurrent. That is, as we see in Figure 175, the lines *AD*, *BE*, and *CF* are concurrent at a point *P*. Truly a fitting Grand Finale to our collection of curiosities!

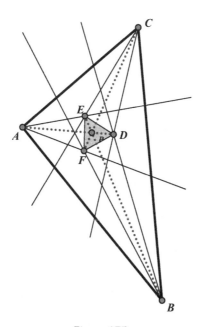

Figure 175

Proofs of the Triangle Curiosities

Curiosity 1. Angles at the Incenter

The basic idea used to prove the relationship that $\angle BIC = 90° + \frac{1}{2}\angle A$ is that the exterior angle of a triangle is equal to the sum of the two remote interior angles.

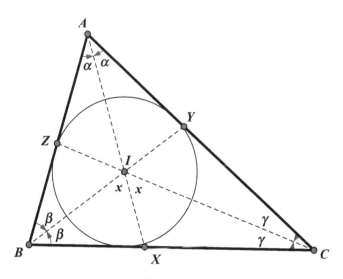

Figure 1-P

In Figure 1-P, we will focus on the exterior angle $\angle BIX$ of triangle AIB where $x = \alpha + \beta$. Similarly, for the exterior angle $\angle XIC$ of triangle AIC, we have $x = \alpha + \gamma$. By addition, $2x = \alpha + \beta + \alpha + \gamma$. Since the sum of the half angles of a triangle is $\frac{1}{2} \cdot 180°$, we then have $2x = \alpha + (\alpha + \beta + \gamma) = \alpha + 90°$. Therefore, $\angle BIC = \frac{1}{2}\angle A + 90°$.

Curiosity 2. An Unexpected Perpendicularity

We begin by considering triangle ABC in Figure 2-P. From Curiosity 1, we know that $\angle AIB = 90° + \frac{1}{2}\angle C$. Since triangle CDE is isosceles, we know that $\angle CDE = \frac{1}{2}(180° - \angle C)$, which leads to $\angle PDB = 90° + \frac{1}{2}\angle C$. Therefore, $\angle AIB = \angle PDB$. Since $\angle PBA = \angle DBP$, triangles AIB and PDC are similar. It follows that $\dfrac{AB}{PB} = \dfrac{BI}{BD}$. This allows us to conclude that triangles APB and IDB are also similar, since $\angle ABP = \angle IBD$, which then allows us to conclude that $\angle APB = \angle IDB$. However, we know that the radius of the inscribed circle is perpendicular to the tangent at the point of tangency. Therefore, $\angle IDB = 90°$, and we can conclude from the triangle similarity that $\angle APB = 90°$, or AP is perpendicular to BP.

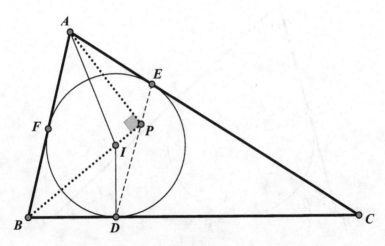

Figure 2-P

Curiosity 3. A Most Unusual Line Bisection

In order to prove this, it will be useful to recall an important property of inscribed and escribed circles[2] of a triangle. Let D be the point at which the inscribed circle of a triangle ABC is tangent to the side BC and G the point at which the escribed circle of ABC opposite A is tangent to BC. Then $BD = CG$, which is shown in Figure 3a-P. This can be shown in the following way.

Since tangents to a circle from an external point are always equal, we can let x denote the lengths of the tangents from A to the incircle,[3] y denote the length of the tangents from B to the incircle, and z denote the lengths of the tangents from C to the incircle. Furthermore, we denote the point at which the excircle is tangent to the extension of AB as Y and the point at which it is tangent to the extension of AC as X. Since $CX = CG$ and $BY = BG$, we have

$$AX + AY = AB + BY + AC + CX = AB + BG + AC + CG$$
$$= AB + BC + CA = 2(x + y + z).$$

Since $AX = AY$, we have $AX = AC + CX = (x + z) + CX = x + y + z$, or $CX = y = BD$. Since $CX = CG$, we therefore have $BD = CG$, as we had set out to show.

With this preparatory background referring to Figure 3a-P, we now let D' denote the point diametrically opposite to point D on the incircle of ABC, as shown in Figure 3b-P, and let G be the point on BC with $DL = LG$. Since L is the midpoint of BC, we have just shown that G is the point of tangency of the excircle of ABC opposite A.

---

[2]An escribed circle of a triangle line is outside the triangle and is tangent to each of the three sides (two of which are extended) of the triangle. This is sometimes referred to as an *excircle*.

[3]Inscribed circles are sometimes referred to as *incircles*.

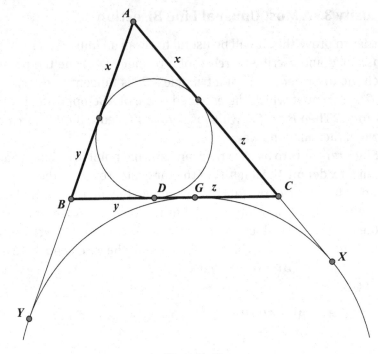

Figure 3a-P

We now consider the homothety[4] with center A that maps the incircle of ABC onto the excircle opposite A. Since the tangent of the incircle in D' is parallel to the tangent at D, that is, perpendicular to BC, this homothety maps D' onto G. It then follows that the center A of the homothety, D', and G are collinear. Since I is the midpoint of DD' and L is the midpoint of DG, we see that LI is parallel to GD'. This means that LZ is parallel to GA, and triangles DLZ and DGA are therefore also

---

[4]Recall that a homothety with center A is a function that maps any point P onto a point Q such that A, P, and Q lie on a common line and $AQ = c \cdot AP$ for some constant value of c. Such a function maps circles onto circles, lines onto parallel lines, and tangents onto tangents. (See Toolbox.)

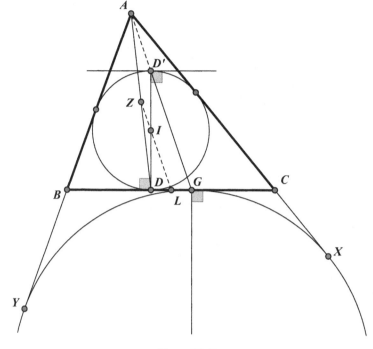

Figure 3b-P

similar. Recalling that L is the midpoint of DG, it follows that Z must be the midpoint of DA, as claimed. It does not make any difference which of the vertices we choose to focus on, as the result we obtain for angle A also holds for each of the other two angles at vertices B and C.

Curiosity 4. Gergonne's Discovery Involving the Inscribed Circle of a Triangle

To prove that the three lines joining vertices to points of tangency of the inscribed circle are concurrent at the Gergonne point, we can use Ceva's theorem (see Toolbox). As a reminder, Ceva's theorem states the following: If three lines are drawn from the vertices of a triangle to the opposite sides creating six segments along the three sides of the triangle, the three lines are concurrent if, and only if, the product of these alternate segments along the sides of the triangle are equal. This is indeed the case for the three segments joining the vertices A, B, and C of the triangle with the points D, E, and F, at which the incircle is tangent to the sides (see Figure 4-P). Since the tangent segments from external points to the same circle are equal, we have $AE = AF$, $BF = BD$, and $CD = CE$. Therefore, $AF \cdot BD \cdot CE = AE \cdot BF \cdot CD$, and the lines AD, BE, and CF are concurrent.

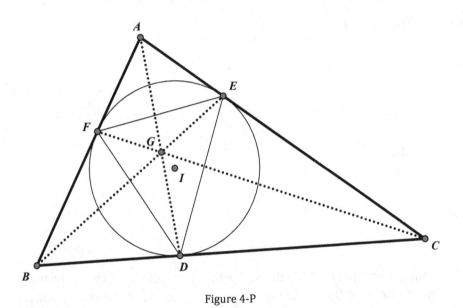

Figure 4-P

Curiosity 5. Another Unexpected Concurrency for Gergonne's Triangle

We notice in Figure 5-P that triangle EAF is isosceles, so that the angle bisector of angle A, which is AU, will be perpendicular to EF. Consider the medial triangle LMN. Since ML is parallel to AB and NL is parallel to AC, the bisector of $\angle MLN$ must be parallel to the bisector of $\angle BAC$. Since LX is parallel to AU, it is therefore the bisector of $\angle MLN$. Similarly, MY is the bisector of $\angle NML$ and NZ is the bisector of $\angle LNM$. This tells us that the three lines LX, MY, and NZ are concurrent, as they are actually the bisectors of the three angles of the medial triangle, which meet at the incenter of the medial triangle.

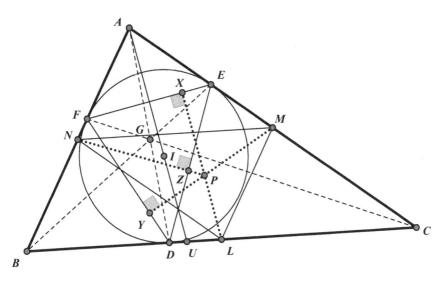

Figure 5-P

Curiosity 6. A Novelty Concerning the Circumscribed and Inscribed Circles

Since points R, S, and T all lie on the circle with center O as shown in Figure 6-P, we can define the radius of this circle as $OR = OS = OT = q$. Similarly, since D, E, and F all lie on the circle with center I, we can also define the radius of this circle as $ID = IE = IF = r$. We now define the point P on the extension of OI beyond I so that $\dfrac{PI}{PO} = \dfrac{r}{q}$ and then consider the triangles PID and POR. Since both ID and OR are perpendicular to BC, we have $\angle PID = \angle POR$. We also know that $\dfrac{PI}{PO} = \dfrac{ID}{OR} = \dfrac{r}{q}$, and triangles PID and POR are therefore similar. This means that P lies on the line RD. Similarly, since P also lies on the lines SE and TF, we see that these three lines have point P in common.

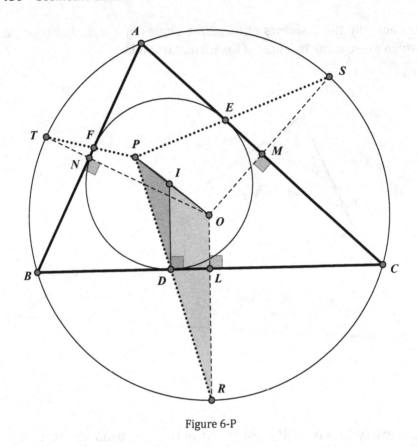

Figure 6-P

Curiosity 7. Concurrency Involving the Inscribed Circle

and

Curiosity 8. A More General Concurrency Involving the Inscribed Circle

Since Curiosity 7 is the special case of Curiosity 8 with $P = I$, the proof for Curiosity 8 also includes the proof for Curiosity 7.

We wish to show that the lines AR, BS, and CT are concurrent. This is equivalent to showing that their poles–with respect to the incircle–are collinear.

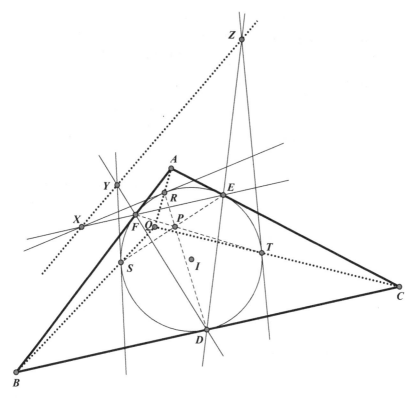

Figure 8-P

The pole[5] of AR is, of course, the common point of the polars of A and R, and as we see in figure 8-P, the polar of A is EF, while the polar of R is the tangent to the incircle at R. We name this point X. Similarly, the pole Y of BS is the common point of FD and the tangent at S, and the pole Z of CT is the common point of DE and the tangent at T. In order

---

[5]Recall that polarity with respect to a circle with center O and radius r is defined as follows. For any point P (other than O) there exists a unique point Q on the ray OP with $OP \cdot OQ = r^2$. The line through Q perpendicular to OP is referred to as the *polar* of P and P is referred to as its *pole*. (See Toolbox.)

to show that X, Y, and Z are collinear, we can apply Menelaus' theorem (see Toolbox) with respect to triangle DEF. Since all three points X, Y, and Z are external to triangle DEF, we can ignore signs and it suffices to show that $\dfrac{EX}{XF} \cdot \dfrac{FY}{YD} \cdot \dfrac{DZ}{ZE} = 1$ holds true.

Considering the first factor of this expression, we note that $\dfrac{EX}{XF} = \left(\dfrac{ER}{RF}\right)^2$. This follows from the law of sines, since we have

$$\frac{EX}{XF} = \frac{EX}{XR} \div \frac{XF}{XR} = \frac{\sin \angle XRE}{\sin \angle REX} \div \frac{\sin \angle XRF}{\sin \angle RFX} = \frac{\sin \angle EFR}{\sin \angle REF} \div \frac{\sin \angle REF}{\sin \angle EFR}$$

$$= \left(\frac{\sin \angle EFR}{\sin \angle REF}\right)^2 = \left(\frac{ER}{RF}\right)^2.$$

Similarly, we also have $\dfrac{FY}{YD} = \left(\dfrac{FS}{SD}\right)^2$ and $\dfrac{DZ}{ZE} = \left(\dfrac{DT}{TE}\right)^2$; therefore,

$$\frac{EX}{XF} \cdot \frac{FY}{YD} \cdot \frac{DZ}{ZE} = \left(\frac{ER}{RF} \cdot \frac{FS}{SD} \cdot \frac{DT}{TE}\right)^2.$$

Since the trigonometric form of Ceva's theorem (see Toolbox) in triangle DEF gives us

$$1 = \frac{\sin \angle EDP}{\sin \angle PDF} \cdot \frac{\sin \angle FEP}{\sin \angle PED} \cdot \frac{\sin \angle DFP}{\sin \angle PFE} = \frac{ER}{RF} \cdot \frac{FS}{SD} \cdot \frac{DT}{TE},$$

the proof is complete.

Curiosity 9. An Inscribed Circle of a General Triangle Generates Equal Angles

In Figure 9-P, since the incenter of a triangle is the point of intersection of its angle bisectors, we know that IC bisects $\angle ACB$. From Curiosity 1, we know that $\angle AIB = 90° + \frac{1}{2} \angle ACB$. Therefore, $\angle BID = 180° - \angle AIB = 90° - \frac{1}{2} \cdot \angle ACB = 90° - \angle ICB = \angle EIC$.

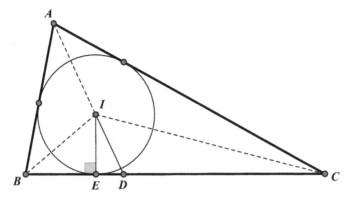

Figure 9-P

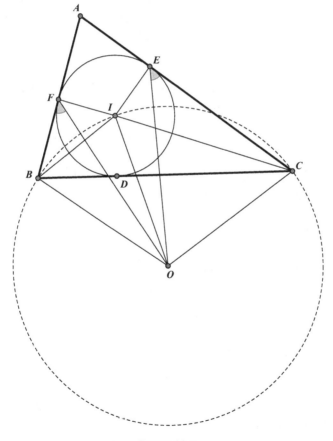

Figure 10-P

Curiosity 10. Surprising Equal Angles

We begin by drawing lines *FI, EI, OI, BO, BI, CI,* and *CO* as shown in figure 10-P. We note that $\angle OIE = \angle OIC + \angle OIE$ and $\angle FIO = \angle FIB + \angle BIO$. Since *I* is the incenter of $\triangle ABC$, we have $\angle CIE = 90° - \angle ECI = 90° - \frac{1}{2}\angle ACB$. Since *OC* = *OI*, triangle *OCI* is isosceles, and we get $\angle OIC = 90° - \frac{1}{2}\angle COI$. However, since $\angle CBI$ is an inscribed angle and $\angle COI$ is a central angle with the same intercepted arc *CI*, we have $\angle COI = 2\angle CBI$. Furthermore, since *BI* is an angle bisector, we have $\angle CBA = 2\angle CBI$. Therefore, $\angle COI = \angle CBA$, and we have $\angle OIC = 90° - \frac{1}{2}\angle CBA$.

Analogously, we can obtain the equations $\angle FIB = 90° - \frac{1}{2}\angle CBA$ and $\angle BIO = 90° - \frac{1}{2}\angle ACB$. Therefore, $\angle OIC + \angle CIE = \angle FIB + \angle BIO$, or $\angle OIE = \angle DIO$. Since *IE* = *IF*, this lets us establish that $\triangle OIF \cong \triangle OEI$; thus, $\angle OFI = \angle IEO$. By subtraction we have $\angle BFO = 90° - \angle OFI = 90° - \angle IEO = \angle OEC$, which is what we set out to prove.

Curiosity 11. An Inscribed Circle of a Triangle Generates its Circumscribed Circle

As we have already seen in Curiosity 1, when *I* is the incenter of triangle *ABC*, as shown in Figure 11-P, we have $\angle BIC = 90° + \frac{1}{2} \cdot \angle BAC$. Since *D* is defined as the intersection of the perpendicular bisectors of *IB* and *IC*, triangles *DIB* and *DCI* are both isosceles, and we have $\angle DBI = \angle BID$ and $\angle DIC = \angle ICD$. Considering the angles in quadrilateral *ICDB*, we can calculate:

$$\angle CDB = 360° - \angle ICD - \angle BIC - \angle DBI$$
$$= 360° - (\angle DIC + \angle BID) - \angle BIC$$
$$= 360° - 2\angle BIC$$
$$= 360° - 2\left(90° + \frac{1}{2}\angle BAC\right)$$
$$= 180° - \angle BAC.$$

We then have point *D* on the circumcircle of *ABC*. We can repeat these calculations in an analogous manner for points *E* and *F* to obtain

$\angle FED = 180° - \angle CBA$ and $\angle DFE = 180° - \angle ACB$. We then see that this is also the case for points E and F, and the points D, E, and F therefore all lie on the circumcircle of triangle ABC.

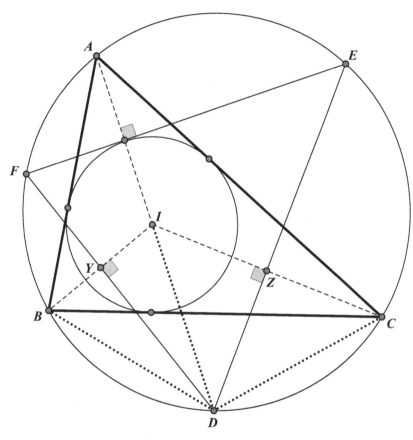

Figure 11-P

Curiosity 12. The Feet of the Altitudes Partition a Triangle into Three Pairs of Equal-Area Triangles

We let X denote the foot of the perpendicular from O to the side BC, as shown in Figure 12-P, and let Y be the foot of the perpendicular from

O to the side *AB*. We can now calculate area[*AOF*] and area[*COD*] in a very straightforward way.

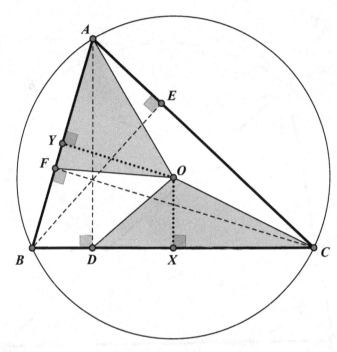

Figure 12-P

The area of triangle *COD* is equal to $\frac{1}{2} \cdot CD \cdot OX$. Since $\angle BOC = 2\angle BAC$ and $\angle XOC = \frac{1}{2} \cdot \angle BOC$, we have $\angle XOC = \angle BAC$. Since the length of *OC* is equal to the radius *r* of the circumcircle, we therefore have $OX = OC \cdot \cos\angle XOC = r \cdot \cos\angle BAC$. Furthermore, in triangle *ACD*, we have $CD = CA = CA \cos\angle ACB$. This gives us

$$\frac{1}{2} \cdot CD \cdot OX = \frac{1}{2} \cdot CA \cdot r \cdot \cos\angle ACB \cdot \cos\angle BAC$$

as the area of triangle *COD*. We can calculate the area of triangle *AOF* in the same way, and this gives us the expression:

$$\frac{1}{2} \cdot AF \cdot OY$$

$$= \frac{1}{2} \cdot \left(CA \cdot \cos \angle BAC\right)\left(OA \cdot \cos \angle ACB\right).$$

$$= \frac{1}{2} \cdot CA \cdot r \cdot \cos \angle ACB \cdot \cos \angle BAC$$

We see that these two expressions are the same, and the areas of the triangles are therefore equal. We can repeat this process for the other two pairs of triangles in exactly the same way, and that will give us area[*BOF*] = area[*COE*] and area[*BOD*] = area[*AOE*].

Curiosity 13. The Noteworthy Position of the Orthocenter of a Triangle

The proof can be done for Figures 13(a) and 13(b) in a similar fashion. We will use Figure 13a-P for simplicity's sake. We begin by showing that $\Delta HBD \sim \Delta HAE$, so that $\dfrac{DH}{EH} = \dfrac{BH}{AH}$ and $AH \cdot DH = BH \cdot EH$. This is true, since both triangles have right angles in points D and E, respectively, and $\angle DAC = \angle BHD = 90° - \angle ACB$. Analogously, we can show that $\Delta HCE \sim \Delta HBF$, so that $\dfrac{CH}{BH} = \dfrac{EH}{FH}$; therefore, $BH \cdot EH = CH \cdot FH$. Consequently, $AH \cdot DH = BH \cdot EH = CH \cdot FH$.

Curiosity 14. A Circle Intersects a Triangle to Generate Lots of Equal Segments

We first note that triangle *HKL* is isosceles since *HK* = *HL*, as shown in Figure 14-P. This means that its altitude from point *H* is the perpendicular bisector of *KL*, and since *NR* is parallel to *BC*, it follows that *KL* is also parallel to *BC*. Furthermore, we see that this perpendicular bisector must pass through *A*, as the line perpendicular to *BC* through *H* is the altitude of triangle *ABC* from vertex *A*. We then have *AK* = *AL*, and analogous arguments also give us

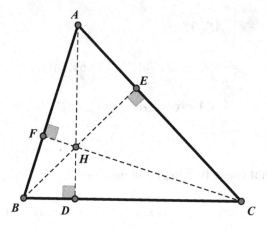

Figure 13a-P

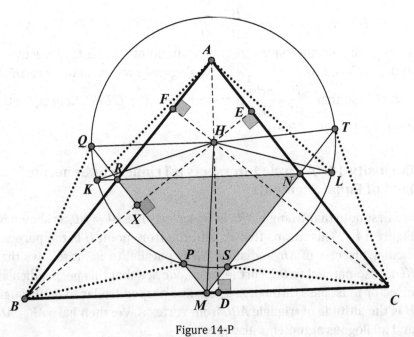

Figure 14-P

$$BP = BQ \text{ and } CS = CT.$$

In quadrilateral $BMHQ$, the diagonals BH and MQ are perpendicular, and we therefore have $BM^2 + HQ^2 = (BX^2 + XM^2) + (HX^2 + XQ^2) = (BX^2 + XQ^2) + (HX^2 + XM^2) = BQ^2 + HM^2$, or $BM^2 + HQ^2 = BQ^2 + HM^2$. Also, in quadrilateral $CTHM$, the diagonals CH and MT are also perpendicular, and we have $CM^2 + HT^2 = CT^2 + HM^2$. Since $BM = CM$ and $HQ = HT$, this gives us $BQ^2 + HM^2 = CT^2 + HM^2$; therefore, $BQ^2 = CT^2$, or $BQ = CT$.

In summary, we now have $BP = BQ = CT = CS$, and an analogous argument for point A then also gives us $AK = AL = BP = BQ = CS = CT$.

Curiosity 15. A Strange Appearance of a Congruent Triangle

The statement of the result was already a bit complicated, and the proof will also prove to require several steps. However, the result is surely worth the effort.

Consider the quadrilateral formed by the points B, C, H, and K in Figure 15-P. Since BK and CH are both perpendicular to the angle bisector EF, they are parallel to each other. As the first (and biggest) step in our proof, we will show that $BCHK$ is a parallelogram, and in order to do that we must further show that $CH = BK$.

As usual, we simplify notation a bit by writing $\angle BAC = \alpha$, $\angle CBA$, $= \beta$ and $\angle ACB = \gamma$, and $BC = a$, $CA = b$, and $AB = c$.

Since D, E, and F are determined by the intersections of the external angle bisectors of ABC, we can calculate the angle measures in triangles BCD, CAE, and ABF and show that these triangles are similar. Consider

$$\angle CAE = \angle BAF = \frac{1}{2}\left(180° - \alpha\right) = 90° - \frac{\alpha}{2},$$

$$\angle ABF = \angle CBD = \frac{1}{2}\left(180° - \beta\right) = 90° - \frac{\beta}{2},$$

and

$$\angle BCD = \angle ACE = \frac{1}{2}\left(180° - \gamma\right) = 90° - \frac{\gamma}{2}.$$

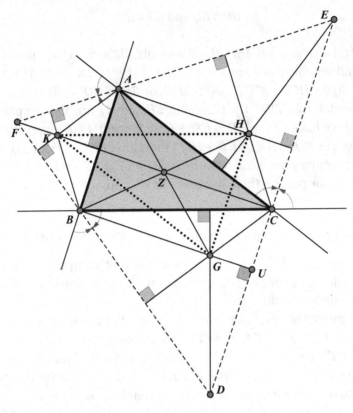

Figure 15-P

The sum of these three angles is $\left(90° - \dfrac{\alpha}{2}\right) + \left(90° - \dfrac{\beta}{2}\right) + \left(90° - \dfrac{\gamma}{2}\right) = 270° - \left(\dfrac{\alpha}{2} + \dfrac{\beta}{2} + \dfrac{\gamma}{2}\right) = 180°.$

The perpendicular from B intersects CD at point U, and then $\angle UBC = 90° - \angle BCD = \dfrac{\gamma}{2}$, and $\angle GCD = 90° - \angle CDB = \dfrac{\alpha}{2}$, and this gives us $CG = a \cdot \sin\dfrac{\alpha}{2} \cdot \sin\dfrac{\gamma}{2}$.

Since triangles BCD and CAE are similar with $\angle BCD = \angle ECA$ and $\angle DBC = \angle AEC$, we have $\dfrac{CH}{CG} = \dfrac{AC}{CD}$; therefore, $CH = \dfrac{CG \cdot AC}{CD}$. The law of sines (see Toolbox) in triangle BCD gives us

$$\frac{CD}{BC} = \frac{\sin \angle DBC}{\sin \angle CDB} = \frac{\sin\left(\dfrac{\alpha}{2} + \dfrac{\gamma}{2}\right)}{\sin\left(\dfrac{\beta}{2} + \dfrac{\gamma}{2}\right)} = \frac{\sin\left(90° - \dfrac{\beta}{2}\right)}{\sin\left(90° - \dfrac{\alpha}{2}\right)} = \frac{\cos\dfrac{\beta}{2}}{\cos\dfrac{\alpha}{2}}.$$

Therefore, $CD = a \cdot \dfrac{\cos\dfrac{\beta}{2}}{\cos\dfrac{\alpha}{2}}$. Together, this gives us

$$CH = \frac{a \cdot \sin\dfrac{\alpha}{2} \cdot \sin\dfrac{\gamma}{2} \cdot b \cdot \cos\dfrac{\alpha}{2}}{a \cdot \cos\dfrac{\beta}{2}} = b \cdot \frac{\sin\dfrac{\alpha}{2} \cdot \sin\dfrac{\gamma}{2} \cdot \cos\dfrac{\alpha}{2}}{\cos\dfrac{\beta}{2}} = \frac{b}{2} \cdot \frac{\sin\alpha \cdot \sin\dfrac{\gamma}{2}}{\cos\dfrac{\beta}{2}}.$$

The analogous calculation on the other side of the figure gives us

$$BK = \frac{a \cdot \sin\dfrac{\alpha}{2} \cdot \sin\dfrac{\beta}{2} \cdot c \cdot \cos\dfrac{\alpha}{2}}{a \cdot \cos\dfrac{\gamma}{2}} = c \cdot \frac{\sin\dfrac{\alpha}{2} \cdot \sin\dfrac{\beta}{2} \cdot \cos\dfrac{\alpha}{2}}{\cos\dfrac{\gamma}{2}} = \frac{c}{2} \cdot \frac{\sin\alpha \cdot \sin\dfrac{\beta}{2}}{\cos\dfrac{\gamma}{2}}.$$

We now need to show that these two values are equal. This follows from the law of sines in triangle *ABC*, since *CH* = *BK* is equivalent to:

$$\frac{b}{2} \cdot \frac{\sin\alpha \cdot \sin\dfrac{\gamma}{2}}{\cos\dfrac{\beta}{2}} = \frac{c}{2} \cdot \frac{\sin\alpha \cdot \sin\dfrac{\beta}{2}}{\cos\dfrac{\gamma}{2}} \quad \text{or} \quad 2b \cdot \sin\dfrac{\gamma}{2} \cdot \cos\dfrac{\gamma}{2} = 2c \cdot \sin\dfrac{\beta}{2} \cdot \cos\dfrac{\beta}{2},$$

which is, in turn, equivalent to $b \cdot \sin \gamma = c \cdot \sin \beta$. We see that *CH* and *BK* are of equal length, and *BCHK* is then a parallelogram, since one pair of opposite sides is both equal and parallel. Now, we are ready for the final argument. We consider the hexagon *AKBGCH*. Since *BCHK* is a parallelogram, its diagonals *BH* and *CK* intersect at their common midpoint. Similarly, we can also show in the same way that *AKGC* (or, if we prefer, *ABGH*) is also a parallelogram whose diagonals *AG* and *CK* intersect at their common midpoint. All three diagonals *AG*, *BH*, and

CK of *AKBGCH* have a common midpoint, which is point *Z*, shown in Figure 15-P.

This means that triangles *ABC* and *GHK* are symmetric with respect to *Z*, and thus are congruent.

Curiosity 16. The Orthocenter Joins Three Other Points in Collinearity

In Figure 16-P, we seek to determine why the orthocenter *H* of triangle *ABC* lies on the line joining *R* and *T*. We will restrict ourselves to showing why this must be the case. A fully analogous argument for line *RS* (or *ST*) can readily be found by relabeling appropriately. Recall that *S* is the common point of *PE* and *AC*. We wish to show that *R*, *S*, *T*, and *H* all lie on a common line, and this is the case if *H* lies both on *RT* and either *RS* or *ST*.

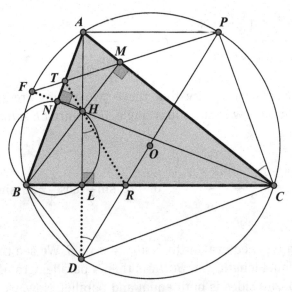

Figure 16-P

We first note that $\angle CDB = 180° - \angle BAC$. Also, considering triangle *BCH*, we have $\angle CBH = \angle CBM = 90° - \angle ACB$ and $\angle HCB = \angle NCB = 90° - \angle CBA$. Thus,

$$\angle BHC = 180° - \angle CBH - \angle HCB = 180° - (90° - \angle ACB) - (90° - \angle CBA)$$
$$= \angle ACB + \angle CBA = 180° - \angle BAC = \angle CDB.$$

Triangles *BCH* and *BDC* are therefore symmetric with respect to *BC*. In other words, we have *LH* = *LD*, and since we also have *RL* ⊥ *DH*, we see that triangle *RHD* is isosceles with ∠*DHR* = ∠*HDR*. We then obtain ∠*DHR* = ∠*HDR* = ∠*ADP* = ∠*ACD*.

Similarly, we also obtain ∠*THF* = ∠*PAC*. We now note that ∠*HNB* = ∠*BLH* = 90°, which means that points *B*, *L*, *H*, and *N* lie on a common circle. From this, we get ∠*NHD* = 180° − ∠*CBA*. Since *P* is a point on the circumscribed circle of *ABC*, we also have ∠*APC* = 180° − ∠*CBA*, which gives us ∠*NHD* = ∠*APC*.

In summary, we have ∠*DHR* + ∠*FHD* + ∠*THF* = ∠*PCA* + ∠*APC* + ∠*CAP* = 180°; therefore, point *H* lies on the line *RT*.

As was stated at the beginning, we can now complete the proof by showing in a completely analogous way that *H* lies either on line *RS* or on line *ST*, and the points *R*, *S*, *T*, and *H* are therefore collinear.

Curiosity 17. The Orthocenter Appears as the Midpoint of a Line Segment

We begin in Figure 17-P by extending *HM* its own length to point *P*. We then have a parallelogram *HBPC*, since the diagonals of this quadrilateral bisect each other. Naming the feet of the altitudes *D* on *BC*, *E* on *CA*, and *F* on *AB*, we now see that ∠*FAE* is supplementary to ∠*EHF*, since the other pair of opposite angles of quadrilateral *AFHE* are both right angles and therefore supplementary. Furthermore, we also have ∠*PBH* supplementary to ∠*BHC* (in parallelogram *HBPC*), which is equal to ∠*EHF*; therefore, ∠*PBH* = ∠*FAE* = ∠*BAC*. We next note that ∠*LKA* = ∠*HKF* = 90° − ∠*FHK* = 90° − ∠*CHL* = ∠*PHC* = ∠*HPB*, and these two angle equalities provide Δ*AKL* ~ Δ *BPH*.

Since ∠*MBH* = 90° − ∠*ACB* = ∠*DAC* = ∠*HAL*, we see that *H* and *M* are points on sides *KL* and *PH* of similar triangles *PHB* and *KLA*, respectively, with ∠*MBH* = ∠*HAL*. We therefore obtain $\dfrac{KH}{LH} = \dfrac{PM}{HM} = 1$, whereupon *KH* = *LH*.

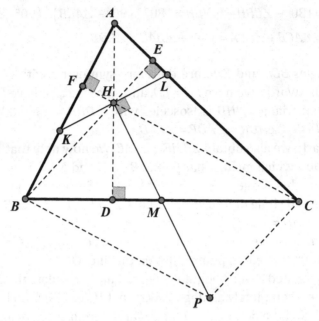

Figure 17-P

Curiosity 18. A Conglomeration of Perpendiculars Generates Equal Line Segments

We begin by applying Menelaus' theorem to triangle *AHC* with transversal *EF* so that we get $\dfrac{AS}{HS}\cdot\dfrac{HF}{CF}\cdot\dfrac{CE}{AE}=1$, (see Toolbox) with point *S* denoting the intersection of *AH* and *EF*, as shown in Figure 18-P. Since *AP* and *RQ* are both perpendicular to *EF*, we find that *AP*∥*RQ*. This means that triangles *DHR* and *DAP* are similar with $\dfrac{HR}{AP}=\dfrac{HD}{AD}$, and we also have similar triangles *HQS* and *APS* so that $\dfrac{HQ}{AP}=\dfrac{HS}{AS}$. Next, we will show that $\dfrac{AD}{HD}\cdot\dfrac{HF}{CF}\cdot\dfrac{CE}{AE}=1$. To simplify matters, we will let *X* = area[*HAB*], *Y* = area[*HBC*], and *Z* = area[*HCA*]. Since two triangles that share the same base have their areas proportional to the lengths of their altitudes to that common base, we get the following relationships: $\dfrac{AD}{HD} = \dfrac{X+Y+Z}{Y}$, $\dfrac{HF}{CF} = \dfrac{X}{X+Y+Z}$, and $\dfrac{CE}{AE} = \dfrac{Y}{X}$.

By multiplying these three equations, we get $\dfrac{AD}{HD} \cdot \dfrac{HF}{CF} \cdot \dfrac{CE}{AE} = 1$, which is what we wanted to prove. Therefore, combining this with the earlier application of Menelaus' theorem, we find that $\dfrac{AS}{HS} = \dfrac{AD}{HD}$. Since we have already shown that $\dfrac{HR}{AP} = \dfrac{HD}{AD}$ and $\dfrac{HQ}{AP} = \dfrac{HS}{AS}$, this gives us $\dfrac{AP}{HQ} = \dfrac{AS}{HS} = \dfrac{AD}{HD} = \dfrac{AP}{HR}$; therefore, $HQ = HR$, which is what we set out to prove.

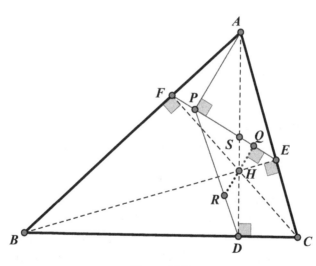

Figure 18-P

Curiosity 19. Yet Another Unexpected Collinearity

Since tangents to a circle from an external point are equal, we have the following equalities: $AE = AF$, $BF = BD$, and $CE = CD$. Therefore, we have $\dfrac{AF}{BF} \cdot \dfrac{BD}{CD} \cdot \dfrac{CE}{AE} = 1$, and by Ceva's theorem (see Toolbox) the lines

AF, *BE*, and *CD* are concurrent, as we see in Figure 19-P. Because these are the lines joining the corresponding vertices of triangle *ABC* and triangle *DEF*, by Desargues's theorem (see Toolbox), the intersections of the corresponding sides, namely, *K*, *L*, and *M*, are collinear.

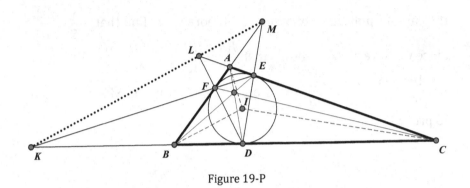

Figure 19-P

Curiosity 20. Introducing the Orthic Triangle

In order to prove that Δ*ABC* ~ Δ*AEF* ~ Δ*EBD* ~ Δ*EFC*, we will prove that Δ*ABC* and Δ*AEF* are similar, and then one can carry out that same procedure for the other triangles Δ*EBD* and Δ*EFC*.

In Figure 20-P, quadrilateral *BFEC* is a cyclic quadrilateral since ∠*BFC* and ∠*BEC* are right angles. Therefore, the angles ∠*CBF* and ∠*FEC* are supplementary. However, ∠*AEF* is supplementary to ∠*FEC*; therefore, ∠*CBF* = ∠*AEF*. Since both Δ*ABC* and Δ*AEF* share ∠*FAE*, we can conclude that Δ*ABC* ~ Δ*AEF*. The same procedure can be followed for the other two triangles, so that we can then conclude that the four triangles are all similar to one another.

To prove that *FC* bisects ∠*DFE*, we again consider the cyclic quadrilateral *BFEC*. In this quadrilateral, we have ∠*CFE* = ∠*CBE* = 90° − ∠*ACB*. Similarly, in the cyclic quadrilateral *CDFA*, we have ∠*DFC* = ∠*DAC* = 90° − ∠*ACB*. We see that ∠*DFC* = ∠*CFE*, and *FC* is therefore the bisector of ∠*DFE*. Similar arguments show us the validity of the claim for the other two lines, *DA* and *EB*.

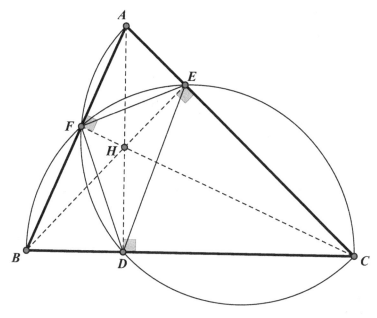

Figure 20-P

Curiosity 21. Finding the Perimeter of the Orthic Triangle

In order to prove this relationship is correct, we consider the points
X and *Y* in Figure 21-P, symmetric to point *D* with respect to *AC* and
AB, respectively. As we have just seen in the proof for Curiosity 20,
we have ∠*FEC* = 180° − ∠*CBF* and ∠*CBF* = ∠*DEC*. This means that
points *E, F,* and *X* are collinear. Similarly, points *E, F,* and *Y* are also
collinear.

Since points *D, P,* and *X* are collinear with *DP* = *PX,* and points *D,*
Q, and *Y* are collinear with *DQ* = *QY,* the homothety (see Toolbox) with
center *D* and factor 2 maps *PQ* onto *XY.* The segment *XY* is therefore
twice as long as *PQ.* Also, since *EX* is symmetric to *ED* with respect to
AC, we have *ED* = *EX,* and, similarly, since *FY* is symmetric to *FD* with
respect to *AB,* we also have *FD* = *FY.* We thus obtain *DE* + *EF* + *FD* =
EX + *EF* + *FY* = *XY* = 2*PQ,* thereby completing the proof.

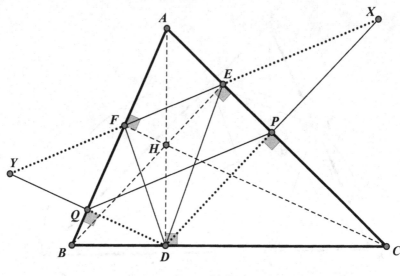

Figure 21-P

Curiosity 22. The Orthic Triangle is a Triangle's Smallest Inscribed Triangle

In Figure 22-P, we assume that *JKL* is a triangle inscribed in the acute triangle *ABC*, with point *J* on side *BC*. Let *X* be the reflection of *J* on *AC* and *Y* the reflection of *J* on *AB*. Because of the resulting symmetry, we have *JK* = *XK* and *LJ* = *LY*. The perimeter of *JKL* is therefore equal to *JK* + *KL* + *LJ* = *XK* + *KL* + *LY*. Since the shortest connection between the points *X* and *Y* is the line segment *XY*, the perimeter of triangle *JKL* is at least as large as the length of *XY*, with the minimum attained when we choose *K* as the common point *K'* of *AC* and *XY* and *L* as the common point *L'* of *AB* and *XY*.

We now note that *AJ* = *AX* = *AY*, because of the symmetry resulting from the reflections. Furthermore, this symmetry also means that

$$\angle YAX = \angle YAB + \angle BAJ + \angle JAC + \angle CAX$$
$$= \angle BAJ + \angle BAJ + \angle JAC + \angle JAC = 2\angle BAC.$$

We see that triangle *AXY* is an isosceles triangle since *AJ* = *AX* = *AY* and ∠*YAX* = 2∠*BAC*, independent of the choice of *J*. The various

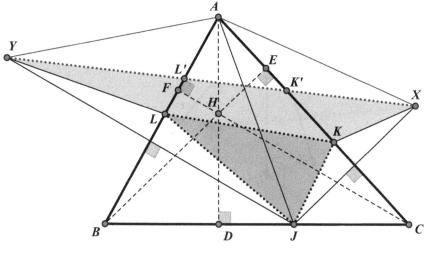

Figure 22-P

possible triangles *AXY* resulting from the different possible choices of *J* on *BC* are therefore all similar, with the smallest perimeter of triangle *JKL* resulting if *AJ* is as small as possible and *K* and *L* are chosen on *AC* and *AB*, respectively. The triangle with the smallest possible perimeter therefore results when *J* is chosen at the foot *D* of the altitude from *A* on *BC*.

This is also true for each of the other two sides, and the triangle *JKL* with minimal perimeter therefore results when *J* overlaps with *D*, *K* with *E*, and *L* with *F*.

Curiosity 23. The Orthic Triangle's Surprising Similar Partner

Before approaching the proof of this curiosity, we recall a useful relationship we have already encountered in the proof of Curiosity 16, namely, that the side of the triangle bisects the section of the perpendicular to the side from the orthocenter to its other endpoint on the circumscribed circle. An alternative way to see this is the following. In Figure 23a-P, the extension of altitude *AD* meets the circumscribed circle at point *K*. By drawing the line *KC*, we have ∠*ABC* = ∠*AKC*, since

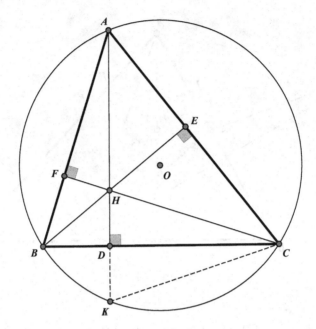

Figure 23a-P

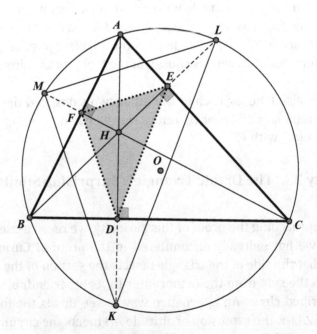

Figure 23b-P

both are measured by the same arc *AC*. Quadrilateral *BFHD* is cyclic since it has a pair of opposite angles that are at right angles, thereby making ∠*FHD* supplementary to ∠*DBF*. However, ∠*DHC* is supplementary to ∠*FHD*. Therefore, ∠*ABC* = ∠*AKC* = ∠*DHC*. Thus, triangle *HKC* is isosceles, with *HD* = *DK*.

Now referring to Figure 23b-P, we have *HD* = *DK*, and analogously *HE* = *EL* and *HF* = *FM*. For triangle *LHM*, we have *EF* as a midline, as it joins the midpoints of two sides of the triangle, and it is therefore parallel to the third side, so that we have *EF*‖*LM*. We can repeat this reasoning to show that *FD*‖*ML* and *DE*‖*KL*. Thus, we have Δ*KLM* ∼ Δ*DEF*.

Curiosity 24. A Nice Concurrency Generated by the Orthic Triangle

As we already know from the proof of Curiosity 20, the altitude *AD* bisects ∠*FDE* in the pedal triangle *DEF*. Similarly, *BE* and *CF* bisect angles ∠*FED* and ∠*DFE*, respectively. In Figure 24-P, we can also note that *MN*‖*DE*, *MR*‖*FD*, and *MJ*‖*AD*. Thus, *MJ* bisects ∠*NMR*. In a similar

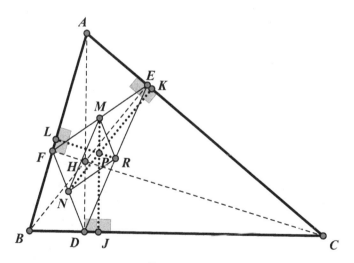

Figure 24-P

fashion, we can show that *NK* bisects ∠*RNM* and that *PL* bisects ∠*MRN*. We then have the three lines *MJ*, *NK*, and *PL* as the angle bisectors of triangle *MNR*, which thus makes them concurrent.

Curiosity 25. Orthic Triangle Generates an Isosceles Triangle

As shown in Figure 25-P, the proof is based on the fact that the circle with diameter *AC* passes through points *D* and *F* because of the right angles ∠*CDA* and ∠*CFA*. Since *M* is the midpoint of *AC*, it is also the center of this circle. We know that the median to the hypotenuse of a right triangle is half the length of the hypotenuse; therefore, in right triangle *ADC*, median *MD* is one-half the length of hypotenuse *AC*. In right triangle *AFC*, median *FM* is also one-half the length of hypotenuse *AC*. Thus, *DM = FM*.

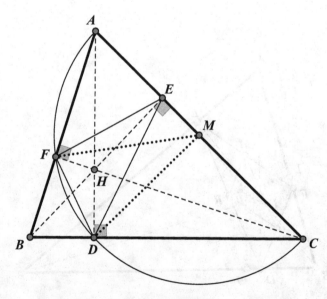

Figure 25-P

Curiosity 26. The Orthic Triangle Generates a Parallelogram

As shown in Figure 26-P, the tangents DL and DM of the inscribed circle of the orthic triangle are of equal length. Thus, triangle DLM is isosceles with altitude DH coincident with AD, and since G is a point on AD, triangle GLM is also isosceles by virtue of the symmetry with respect to AD.

To prove that $LJEG$ is a parallelogram, we will make use of the many right triangles in the configuration. Consider the following: $\angle BAC = \alpha$, $\angle CBA = \beta$, $\angle ACB = \gamma$, $BC = a$, $CA = b$, and $AB = c$. We then have in triangle EBC that $EB = c \sin \angle BAE = c \sin \alpha$, and, where $EY \perp BC$, in triangle EBY that $EY = EB \sin \angle YBE = c \sin \alpha \sin (90° - \gamma) = c \sin \alpha \cos \gamma$.

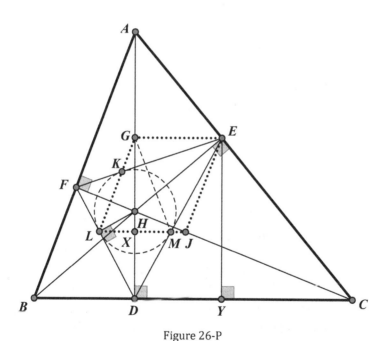

Figure 26-P

We will now show that $DG = EY$ in order to establish that EG is parallel to BC. Noting that LM is also parallel to BC since DH is perpendicular to *BC* and the base *LM* of the isosceles triangle *DLM* is also perpendicular to *DH*, we see that *EG* is then parallel to *LJ*. An analogous argument then shows us that *GL* is parallel to *EJ*, and quadrilateral *EGLJ* is thus a parallelogram.

In order to calculate the length of *DG*, we note that $DG = DX + XG$ holds. In triangles *BDA*, *DBH*, *HDL*, and *XDL*, respectively, we have

$$BD = c \cos \angle DBA = c \cos \beta,$$

$$DH = BD \tan \angle DBH = c \cos \beta \tan (90° - \gamma) = c \cos \beta \cot \gamma,$$

$$DL = DH \sin \angle HDL = c \cos \beta \cot \gamma \cos (90° - \alpha) = c \cos \beta \cot \gamma \sin \alpha,$$

$$DX = DL \cos \angle XDL = c \cos \beta \cot \gamma \sin \alpha \cos (90° - \alpha)$$
$$= c \cos \beta \cot \gamma \sin^2 \alpha,$$

and in triangles *XDL* and *LGX* we have

$$LX = DL \sin \angle XDL = c \cos \beta \cot \gamma \sin \alpha \sin (90° - \alpha)$$
$$= c \cos \beta \cot \gamma \sin \alpha \cos \alpha,$$

$$XG = LX \cot \angle LGX = c \cos \beta \cot \gamma \sin \alpha \cos \alpha \cot (90° - \beta)$$
$$= c \cos (\beta) \cot \gamma \sin \alpha \cos \alpha \tan \beta$$

$$= c \sin \beta \cot \gamma \sin \alpha \cos \alpha.$$

Summing up, this gives us

$$DG = DX + XG = c \cos \beta \cot \gamma \sin^2 \alpha + c \sin \beta \cot \gamma \sin \alpha \cos \alpha$$

$$= c \cot \gamma \sin \alpha (\cos \beta \sin \alpha + \sin \beta \cos \alpha) = c \cot \gamma \sin \alpha \sin (\alpha + \beta)$$
$$= c \cot \gamma \sin \alpha \sin (180° - \gamma) = c \cot \gamma \sin \alpha \sin \gamma = c \sin \alpha \cos \gamma,$$

which is equal to the previously calculated length of *EY*. As stated above, this means that both *EG* and *LJ* are parallel to *BC*; they are therefore also parallel to each other. Since we can repeat these calculations to show that both *EJ* and *GL* are parallel to *AB*, this establishes that quadrilateral *EGLJ* is a parallelogram.

Curiosity 27. Concyclic Points Generated by the Orthic Triangle

This proof can benefit from our findings in Curiosity 26, where we have ∠*HKE* = ∠*EMH* = 90°, and points *K* and *M* certainly lie on the circle with diameter *EH*, whose center, *O*, is the midpoint of *EH*. We see this again in Figure 27-P. Since *LM* (or *LJ*) is parallel to *BC*, we have ∠*HJM* = ∠*FJL* = ∠*FCB* = 90° – β, and from Curiosity 20, we already know that ∠*HEM* = 90° – β. We therefore have ∠*HJM* = 90° – β = ∠*HEM*, and point *J* also lies on the circle with diameter *EH*. Since an analogous argument can also be made for point *G* on the other side, we have all six points, namely, points *H*, *M*, *J*, *E*, *G*, and *K*, lying on the same circle.

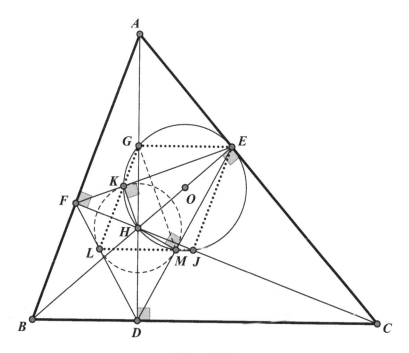

Figure 27-P

Curiosity 28. The Orthic Triangle Generates Concurrent Lines

Let Q' denote the common point of lines EF and BC and let X denote the common point of EF and AD, as shown in Figure 28-P. We note that we have $A = FB \cap EC$, $H = FC \cap EB$, $Q' = BC \cap EF$, and $X = AH \cap EF$. By applying Ceva's theorem (see Toolbox) to triangle AEF with respect to point H, we obtain $\dfrac{EX}{XF} \cdot \dfrac{FB}{BA} \cdot \dfrac{AC}{CE} = 1$. Ignoring the orientation of the line segments, by applying Menelaus' theorem (see Toolbox) to triangle AEF with respect to line BC, we obtain $\dfrac{EQ'}{Q'F} \cdot \dfrac{FB}{BA} \cdot \dfrac{AC}{CE} = 1$. Combining these two results, we therefore have $\dfrac{EX}{XF} = \dfrac{EQ'}{Q'F}$.

If we similarly let Q'' denote the common point of lines EF and GJ, we note $FG \cap EJ = D$, $JF \cap EG = P$, and $EF \cap JG = Q''$. We can now apply Ceva's theorem to triangle DEF with respect to point P to obtain $\dfrac{EX}{XF} \cdot \dfrac{FG}{GD} \cdot \dfrac{DJ}{JE} = 1$, and apply Menelaus' theorem to triangle DEF with respect to line GJ to obtain $\dfrac{EQ''}{Q''F} \cdot \dfrac{FG}{GD} \cdot \dfrac{DJ}{JE} = 1$. Combining these two results gives us $\dfrac{EX}{XF} = \dfrac{EQ''}{Q''F}$. It therefore follows that $Q' = Q'' = Q$ is a common point of BC, EF, and GJ, as claimed.

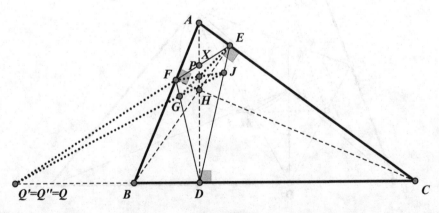

Figure 28-P

Curiosity 29. The Orthic Triangle Generates Collinear Points

This is a special case of a situation usually referred to as a *trilinear polarity*. Such a situation results when we intersect the sides of a triangle with the lines joining the opposite vertices with some common interior point, the point here (illustrated in Figure 29-P) being the orthocenter *H*, which lies on the lines *AD*, *BE*, and *CF*. We now consider the two triangles *ABC* and *DEF*. It is an immediate consequence of Desargues's theorem (see Toolbox) that the points *K* = *BC* ∩ *EF*, *L* = *CA* ∩ *FD*, and *M* = *AB* ∩ *DE* lie on a common line.

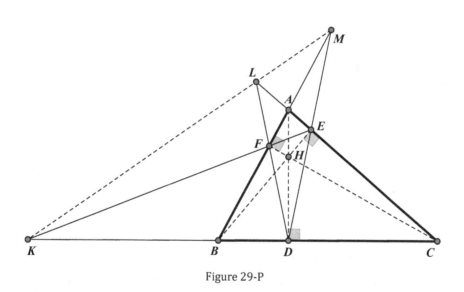

Figure 29-P

Curiosity 30. An Unexpected Property of Altitude Feet of a Triangle

As we see in Figure 30-P, lines *EM* and *FM* are medians of right triangles *BCE* and *BCF* to their common hypotenuse *BC*. We know that the median to the hypotenuse of a right triangle is half the length of the hypotenuse, and we therefore have *ME* = *MF*. We note that ∠*A* is a common angle of the right triangles *ABE* and *AFC*. This gives us ∠*EBA* = 90° − ∠*A* = ∠*ACF*. Finally, we know that *ME* = *MF*. Triangle *MEF* is then isosceles, and the perpendicular bisector of its base *EF* thus contains the third vertex, namely, point *M*.

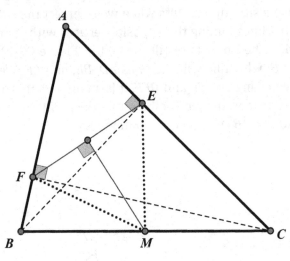

Figure 30-P

Curiosity 31. How a Non-Isosceles Triangle Can Generate a Parallelogram

In Figure 31-P, we begin with $\angle AOB = 2\angle ACB$.

Also, $\angle BAO = \frac{1}{2}(180° - \angle AOB) = \frac{1}{2}(180° - 2\angle ACB) = 90° - \angle ACB$.

Furthermore, $\angle GAC = 90° - \angle ACB$. Therefore, $\triangle AED \sim \triangle AGC$ and $\frac{AE}{AD} = \frac{AG}{AC}$. By addition of $\angle DAG$, we also have $\angle BAG = \angle DAC$, and from this we have $\triangle AEG \sim \triangle ADC$. Therefore, $\angle AEG = \angle ADC$. Since $\triangle ABC$ is not isosceles, we have $BD \ne DC$ (if AD were the perpendicular bisector of BC, then $AB = AC$ would follow). Since $OB = OC$ and $\angle ODB = \angle CDO$, O lies on the circumcircle of BDC. Quadrilateral $OBDC$ is therefore a cyclic quadrilateral, and we have $\angle CBO = \angle CDA = \angle GEA$. Since O is the center of the circumcircle of $\triangle ABC$, we have $\angle OBC + \angle BAC = 90°$, and we then have $\angle GEA + \angle BAC = 90°$. Thus, $EG \perp AG$, and $EG \| DF$. An analogous argument also gives us $GF \| ED$, and $DFGE$ is therefore a parallelogram.

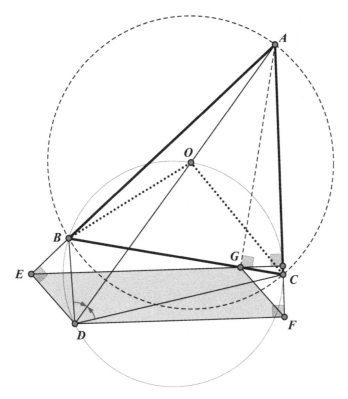

Figure 31-P

Curiosity 32. Perpendiculars Generating Unexpected Parallel Lines

As a first step, we add a few useful lines to Figure 32-P, namely, PA, PC, and CD. With the aid of these lines, we will show $\angle FCA = \angle CAD$, and, therefore, $FC \| AD$. Since A, F, P, and C all lie on a common circle, we have $\angle FCA = \angle FPA$, as they are measured by the same arc FA. In triangle LAP, we obtain the following:

$$\angle FPA = \angle LPA = 90° - \angle PAL = 90° - (180° - \angle BAC - \angle CAP)$$
$$= \angle BAC + \angle CAP - 90°.$$

Similarly, since D, A, P, and C all lie on a common circle, using a common arc measure, we also have $\angle DAC = \angle DPC$ and $\angle CDP =$

$\angle CAP$. Furthermore, since triangle *JDC* is a right triangle and *ABCP* an inscribed quadrilateral, it follows that $\angle PCD = \angle PCB + \angle BCD = (180° - \angle BAP) + \angle JCD = (180° - \angle BAC - \angle CAP) + (90° - \angle CDJ) = (180° - \angle BAC - \angle CAP) + (90° - \angle CDP) = (180° - \angle BAC - \angle CAP) + (90° - \angle CAP) = 270° - \angle BAC - 2 \cdot \angle CAP$

Now, considering triangle *PDC*, we have the following:

$$\angle DPC = 180° - \angle CDP - \angle PCD = 180° - \angle CAP$$
$$- (270° - \angle BAC - 2\angle CAP) = \angle BAC + \angle CAP - 90°.$$

Summing up, we have $\angle DAC = \angle DPC = \angle BAC + \angle CAP - 90° = \angle FPA = \angle FCA$, and thus, *FC∥AD*. Analogously, since we can also show that *BE∥AD*, we can see that the three lines *FC*, *AD*, and *BE* are indeed parallel.

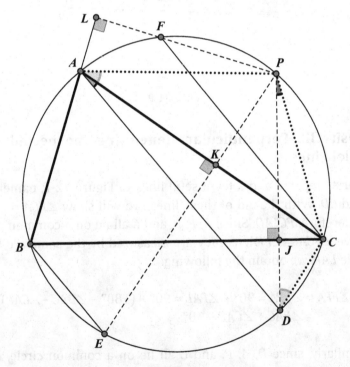

Figure 32-P

Curiosity 33. Tangents and Inscribed-Triangle Sides Generate a Surprising Collinearity

In Figure 33-P, we note that triangles AEB and BEC are similar. This is because they share an angle in E and $\angle ECB = \angle ABE$, since both are measured by one-half arc AB. This enables us to conclude that $\dfrac{AE}{BE} = \dfrac{BE}{CE} = \dfrac{AB}{BC} = \dfrac{x}{y}$, and from this we obtain $\dfrac{AE}{CE} = \dfrac{AE}{BE} \cdot \dfrac{BE}{CE} = \dfrac{x^2}{y^2} = \left(\dfrac{AB}{BC}\right)^2$. Analogously, just as point E partitions line CE proportionally, the same is true for points D and F on lines BD and AF, respectively, so that $\dfrac{CD}{BD} = \left(\dfrac{CA}{AB}\right)^2 = \dfrac{z^2}{x^2}$, and $\dfrac{BF}{FA} = \left(\dfrac{BC}{CA}\right)^2 = \dfrac{y^2}{z^2}$. Thus, the product of

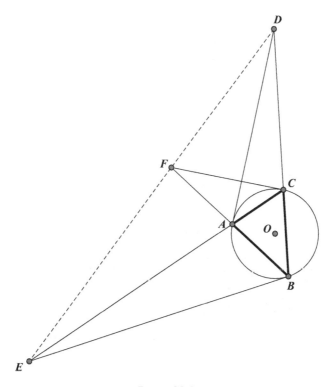

Figure 33-P

these ratios into which the points D, E, and F partition the sides of triangle ABC gives us $\dfrac{AE}{EC} \cdot \dfrac{CD}{DB} \cdot \dfrac{BF}{FA} = \dfrac{x^2}{y^2} \cdot \dfrac{z^2}{x^2} \cdot \dfrac{y^2}{z^2} = 1$, which, according to Menelaus' theorem (see Toolbox), indicates that the points D, E, and F are collinear.

Curiosity 34. Creating Congruent Triangles Inscribed in the Same Circle

Surprisingly, this is an immediate consequence of Curiosity 32. Since FC, AD, and BE are parallel chords of circle O, their perpendicular bisectors are identical and pass through the center of circle O, as we see in Figure 34-P. Since D, E, and F lie symmetric to A, B, and C, respectively, with respect to this common bisector (see Toolbox), the triangles ABC and DFE are symmetric and therefore congruent.

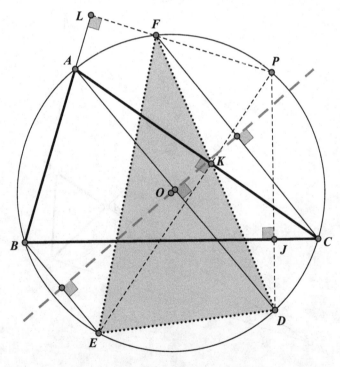

Figure 34-P

Curiosity 35. An Unexpected Product Equality

We begin in Figure 35-P by drawing line segment *CE*. We then have right angle *ECB*, which is inscribed in the semicircle. The two right triangles *ABD* and *EBC* are similar, since both ∠*BAC* and ∠*BEC* are measured by the same intercepted arc *BC*. This enables us to set up the proportion $\dfrac{AB}{BE} = \dfrac{BD}{BC}$, which leads to our desired conclusion, $AB \cdot BC = BD \cdot BE$.

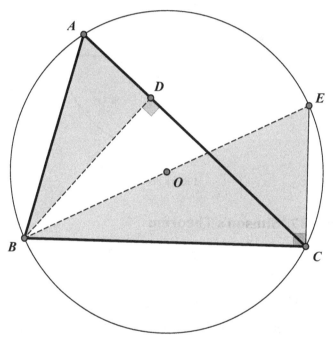

Figure 35-P

Curiosity 36. An Unusual Product of Two Sides of a Triangle

The proof of this is a rather simple application of similar triangles. We begin in Figure 36-P by drawing *CD* to create triangle *ADC*. Since ∠*ACD* is inscribed in a semicircle, it is a right angle. Also, ∠*EBA* and ∠*CDA* are both inscribed angles with the same intercepted arc *AC*.

Therefore, $\angle EBA = \angle CDA$, thus making $\triangle ABE \sim \triangle ADC$. This enables us to get $\dfrac{AB}{AD} = \dfrac{AE}{AC}$, and then $AD \cdot AE = AB \cdot AC$.

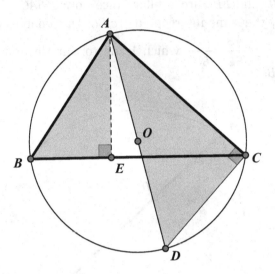

Figure 36-P

Curiosity 37. Simson's Theorem

We begin by drawing PA, PB, and PC in Figure 37-P. Since A, B, C, and P lie on a common circle, we have $\angle ABP = \angle ACP$. Right triangles ZPB and YPC are therefore similar, and we have $\dfrac{BZ}{CY} = \dfrac{PZ}{PY}$. Similarly, $\angle PAB = \angle PCB$ gives us similar right triangles XCP and ZAP, so that

$$\frac{CX}{AZ} = \frac{PX}{PZ}.$$

Since $\angle CBP$ and $\angle PAC$ are opposite angles of an inscribed (cyclic) quadrilateral, they are supplementary. However, $\angle YAP$ is also supplementary to $\angle PAC$. This means that right triangles XPB and YPA are also similar, giving us $\dfrac{AY}{BX} = \dfrac{PY}{PX}$. Multiplying these identities, we obtain $\dfrac{BZ}{CY}\cdot\dfrac{CX}{AZ}\cdot\dfrac{AY}{BX}=\dfrac{PZ}{PY}\cdot\dfrac{PX}{PZ}\cdot\dfrac{PY}{PX}=1$. Thus, by Menelaus' theorem (see Toolbox), the points X, Y, and Z are collinear. These three points determine the *Simson line* of $\triangle ABC$ with respect to point P.

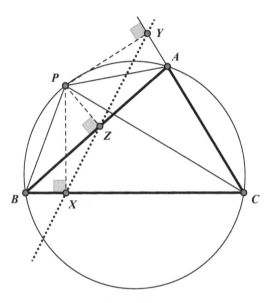

Figure 37-P

Curiosity 38. An Extension of Simson's Theorem

We know that in Figure 38-P the altitude extended in triangle ABC from vertex C to side AB meets the circumcircle at point P and meets AB at point Z, which is one of the points on the Simson line. We also have PX and PY perpendicular, respectively, to sides BC and CA of triangle ABC. Therefore, points X, Y, and Z determine the Simson line of P with respect to triangle ABC.

We now draw PA and consider quadrilateral $PXAY$, where $\angle AZP = \angle PYA = 90°$. This makes $PXAY$ a cyclic quadrilateral, and we therefore have $\angle ZYA = \angle ZPA$. We now let E denote a point on the tangent of the circumcircle of ABC at point C, as shown in Figure 38-P. In this circle, we certainly have $\angle ECA = \angle CPA$, since both angles are measured by one-half measure of arc *AC*. We then obtain $\angle ZYC = \angle ZYA = \angle ZPA = \angle CPA = \angle ECA$. Since they are the alternate-interior angles of the Simson line *XYZ* and the tangent line *EC*, these two lines are parallel.

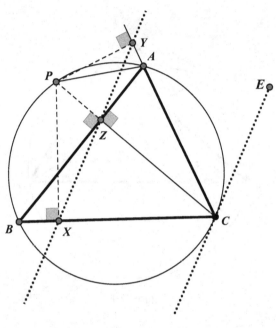

Figure 38-P

Curiosity 39. An Interesting Aspect of Simson's Theorem

In Figure 39-P, we extend altitude *CH* to meet the circumcircle of *ABC* at point *T* and extend *PZ* to meet the circumcircle at point *K*. After constructing *CK*, we draw the line from point *H* parallel to *CK*. This line meets *PK* at point *U*. We then have parallelogram *CKUH*, whereupon *HU* = *CK*. We also have isosceles trapezoid *PKCT*, so that *PT* = *CK*. Therefore, *TPUH* is also an isosceles trapezoid. We now consider triangle *ATH*. Since *T* is a point on the circumcircle of triangle *ABC*, we have $\angle HTA = \angle CTA = \angle CBA$. Also, we have $\angle AHE = \angle CBA$, since their respective sides are perpendicular, as $AH \perp BC$ and $EH \perp BA$. Therefore, triangle *ATH* is isosceles. Letting *E* denote the midpoint of *TH*, we see that line *AB* (or *AE*) is the perpendicular bisector of *TH*. Since point *Z* lies on this line, and recalling that *TPUH* is an isosceles trapezoid, we see that *Z* is the midpoint of *PU*. Now, we must show that *HU* is parallel to the Simson line *XYZ*. In order to do this, we note that quadrilateral *PZAY* is cyclic, since we have $\angle PYA = \angle AZP = 90°$. From this, we obtain $\angle ZYC = \angle ZYA = \angle ZPA = \angle KPA$, and since *A*, *P*, *K*, and *C* lie on a common circle, we have $\angle KPA = 180° - \angle ACK = 180° - \angle YCK$. Summarizing, this gives us $\angle ZYC = 180° - \angle YCK$, and *CK* is therefore parallel to the Simson line *YZ*. Since we know that *HU* is parallel to *CK*, it is also parallel to the Simson line.

As a last step, we now consider triangle *PUH*. Since the Simson line *ZR* is parallel to side *HU* of triangle *PUH*, and *Z* is the midpoint of its side *PU*, it must also intersect the other the side of triangle *PUH* at its midpoint, *R*. Therefore *PR* = *HR*, and *R* is the midpoint of *PH* as we had set out to prove.

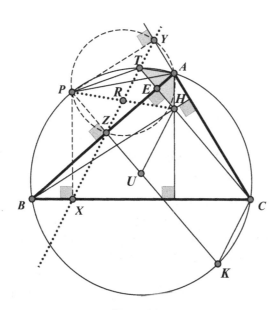

Figure 39-P

Curiosity 40. A Parallel to the Simson Line

In Figure 40-P, we see a slightly different version of the same constellation. Since P and Q both lie on the circumcircle of ABC, we have $\angle AQP = \angle ABP$. Noting that $\angle PXB = \angle PZB = 90°$, points P, B, X, and Z also lie on a common circle, and we have $\angle ZXP = \angle ZBP$. Thus, we have $\angle ZXP = \angle ZBP = \angle ABP = \angle AQP$, and since AQ and XZ intersect PXQ, they form equal corresponding angles $\angle AQX = \angle YXP$, and we have $AQ \| YX$.

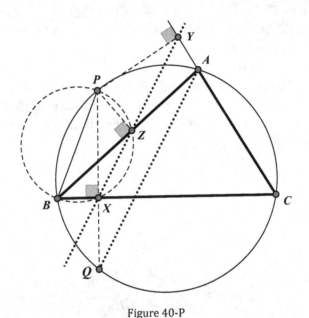

Figure 40-P

Curiosity 41. Two Triangles Related by a Common Point: Circumcenter – Centroid

In Figure 41-P, we first note that X is the point of intersection of the perpendicular bisectors of BG and CG. This means that segments BX, CX, and GX are all the same length, and point X is the circumcenter of triangle BCG. It therefore also lies on the bisector of BC, namely, the line OL. Similarly, Y lies on OM, and Z lies on ON. We now notice some angle equalities. Since AQ is perpendicular to QY, and AM is perpendicular to MY, points A, Q, M, and Y are concyclic, and we have $\angle QAM = \angle QYM$, or $\angle LAC = \angle DYO$. Similarly, points C, L, O, and M are also concyclic. Defining the point D, where the extension of LO meets YZ, we then obtain $\angle MCL = \angle TOD$, since both angles are supplementary to $\angle LOM$. Since $\angle MCL = \angle ACL$, we have $\triangle YOD \sim \triangle ACL$, and we get $\dfrac{YD}{AL} = \dfrac{OD}{CL}$.

Analogously, we can find that $\triangle ABL \sim \triangle ZOD$, which gives us $\dfrac{ZD}{AL} = \dfrac{OD}{BL}$. Since $BL = CL$, we can conclude that $YD = ZD$, which implies that the

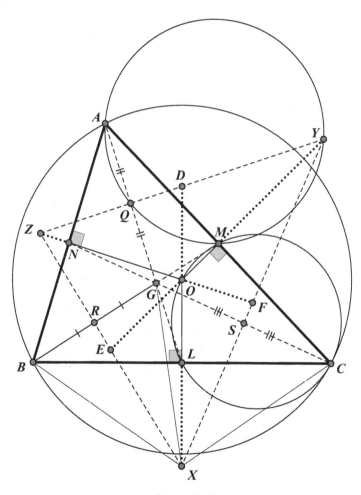

Figure 41-P

point D, which is where XO extended meets YZ, is the midpoint of YZ. This can be then replicated for the other two midpoints, E and F, which allows us to conclude that point O is the centroid of $\triangle XYZ$.

Curiosity 42. Introducing the Medians of a Triangle with Some of Their Amazing Properties

The median of a triangle partitions the triangle into two equal area triangles, since the bases of the two triangles are equal and they share the same altitude, as is the case with area[ABD] = area[ADC] in Figure 42a-P. In Figure 42b-P, using the same argument, we can show that median GD partitions triangle BGC into two equal parts, namely, area[GBD] = area[GCD]. From this, we see that

$$\begin{aligned} \text{area}[ABG] &= \text{area}[ABD] - \text{area}[GBD] \\ &= \text{area}[ADC] - \text{area}[GDC] = \text{area}[AGC]. \end{aligned}$$

Since area[GAF] = area[GFB] and area[GCE] = area[GEA], we obtain $\frac{1}{2}$ area[ABG] = $\frac{1}{2}$ area[AGC], which gives us area[GAF] = area[GFB] = area[GCE] = area[GEA].

When we repeat this for triangle BCE and triangle BEA, we find equality for the six triangles shown, namely, that

$$\begin{aligned} \text{area}[GBD] &= \text{area}[GDC] = \text{area}[GCE] \\ &= \text{area}[GEA] = \text{area}[GAF] = \text{area}[GFB]. \end{aligned}$$

Since area[GDC] = area[GCE] = area[GEA], we therefore have area[GDC] = $\frac{1}{3}$ area[ADC]. The altitude of triangle GDC on base CD is therefore one third the altitude of ADC, from which we also see that GD is one third of AD. Point G is therefore a trisection point of median AD, and since we can argue in an analogous way for the other two medians, it is also a trisection point of BE and CF.

In order to see why the shortest median has its endpoint on the longest side of the triangle, we find that it will be useful to apply Apollonius' theorem (see Toolbox). This states that

$$AB^2 + AC^2 = 2(AD^2 + BD^2)$$

in any triangle *ABC*, where *D* is the midpoint of side *BC*. This is an immediate consequence, illustrated in Figure 42a-P, of the law of cosines, since we have $AB^2 = AD^2 + BD^2 - 2AD \cdot BD \cdot \cos\angle ADB$ and

$$AC^2 = AD^2 + CD^2 - 2AD \cdot CD \cdot \cos\angle ADC$$
$$= AD^2 + BD^2 - 2AD \cdot BD \cdot \cos(180° - \angle ADB)$$
$$= AD^2 + BD^2 + 2AD \cdot BD \cdot \cos\angle ADB.$$

Adding these two equations gives us $AB^2 + AC^2 = 2(AD^2 + BD^2)$.

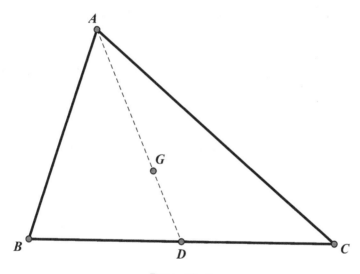

Figure 42a-P

Having established this and noting $BD = \frac{1}{2}BC$, we can also write Apollonius' theorem as

$$AD^2 = \frac{1}{4}\left(2AB^2 + 2AC^2 - BC^2\right).$$

We can now compare the lengths of any two medians, and we see that $AD > BE$ is equivalent to

$$\frac{1}{4}\left(2AB^2 + 2AC^2 - BC^2\right) > \frac{1}{4}\left(2AB^2 + 2BC^2 - AC^2\right), \text{ or } 3AC^2 > 3BC^2.$$

We see that the longer median has its endpoint on the shorter side of *ABC*, and the shortest median, therefore, has its endpoint on the longest side of the triangle, as claimed. In order to see that the sum of the lengths of the medians is less than the perimeter of the triangle, we note that the sum of the lengths of any two sides of a triangle is always greater than the length of the third side. This is because the third side joins two vertices in the shortest path between them, which the other two points join by an indirect route through the third vertex. In Figure 42a-P, this means that $AB + BD > AD$, $FB + BC > CF$, $BC + CE > BE$, $DC + CA > AD$, $CA + AF > CF$, and $EA + AB > BF$. Adding all of these inequalities gives us $2(AB + BC + CA) > 2(AD + BE + CF)$, or $AB + BC + CA > AD + BE + CF$.

Next, we show that, for any triangle, the sum of the lengths of the medians is greater than three-fourths of the perimeter of the triangle. We begin by using the well-known trisection property of the centroid *G* of △*ABC* on each median, and, for simplicity's sake, we will refer to the medians from vertices *A*, *B*, and *C* as m_a, m_b, and m_c.

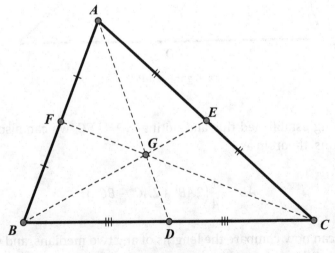

Figure 42b-P

As shown in Figure 42b-P, in $\triangle BGC$, $BG + CG > BC$, or $\frac{2}{3}m_c + \frac{2}{3}m_b > a$. In a similar way we get $\frac{2}{3}m_b + \frac{2}{3}m_a > c$ and $\frac{2}{3}m_a + \frac{2}{3}m_c > b$.

By addition: $\frac{4}{3}(m_a + m_b + m_c) > a + b + c$, and therefore, $m_a + m_b + m_c > \frac{3}{4}(a + b + c)$.

Finally, we show $AD^2 + BE^2 + CF^2 = \frac{3}{4}(AB^2 + BC^2 + CA^2)$ by again applying Apollonius' theorem. We know that $AD^2 = \frac{1}{4}(2AB^2 + 2CA^2 - BC^2)$ holds, and similarly, we also have $BE^2 = \frac{1}{4}(2AB^2 + 2BC^2 - CA^2)$ and $CF^2 = \frac{1}{4}(2BC^2 + 2CA^2 - AB^2)$. Adding these three equations gives us the required result: $AD^2 + BE^2 + CF^2 = \frac{3}{4}(AB^2 + BC^2 + CA^2)$.

Curiosity 43. The Median of a Triangle is Equidistant from Two Vertices

The proof, shown in Figure 43-P, is rather simple, yet the result is rather noteworthy. Since $BD = CD$ and $\angle BDP = \angle QDC$, we can establish the congruent right triangles $\triangle CQD \cong \triangle BFD$. Therefore, $BP = CQ$.

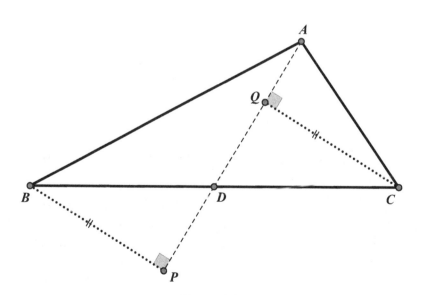

Figure 43-P

Curiosity 44. The Special Median of a Right Triangle

A simple way to prove this relationship is to consider that the right triangle inscribed in a circle has its hypotenuse as the diameter of the circle, as shown in Figure 44-P. Obviously, the midpoint, *M*, of the hypotenuse *BC* of triangle *ABC* is also the center of the circumscribed circle, where *MA*, *MB*, and *MC* are the radii, and therefore, $AM = \frac{1}{2}BC$.

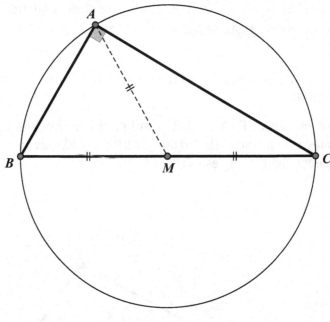

Figure 44-P

Curiosity 45. Medians Partition Any Triangle into Four Congruent Triangles

The bisector property of point *L* follows immediately from the fact that there exists a homothety (see Toolbox) with center *A* and factor 2 that maps triangle *AFE* onto triangle *ABC*, combined with the fact *D* is the midpoint of *BC*.

To prove the congruence of the four small triangles can be done by recognizing that the midlines of the triangle are half the length of the side to which they are parallel. We see this in Figure 45-P. Therefore, for example, $EF = \frac{1}{2}BC = BD$. The same argument can be made for $FD = AF$, and we know that since F is the midpoint of AB, we have $AF = FB$. We have thus shown that $\triangle AFE \cong \triangle FBD$. In a similar manner we can show that $\triangle ADF \cong \triangle DFE \cong \triangle FEC \cong \triangle DEB$.

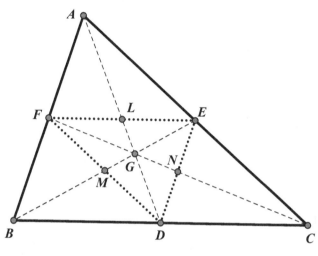

Figure 45-P

Curiosity 46. How the Centroid Helps Create a Similar Triangle

Since GK is parallel to BC, $\angle GKH = \angle CBH$, and $\angle HGK = \angle HCB$, we note that triangles HKG and HBC in Figure 46-P are similar. Triangles AGK and ADC are also similar, and since the centroid G partitions a median into thirds, we have $\dfrac{GK}{DC} = \dfrac{AG}{AD} = \dfrac{2}{3}$ and, therefore, $\dfrac{GH}{HC} = \dfrac{GK}{BC} = 2\left(\dfrac{GK}{DC}\right) = 2\left(\dfrac{2}{3}\right) = \dfrac{1}{3}$. Thus, we have $\dfrac{CH}{CF} = \dfrac{CH}{CG} \cdot \dfrac{CG}{CF} = \dfrac{3}{4} \cdot \dfrac{2}{3} = \dfrac{1}{2}$, which establishes that H is the midpoint of line segment CF. Since D is the midpoint of BC, we therefore obtain $HD \| AB$. Analogously, we also have $JD \| AC$, which we can show by considering similar triangles AGL and ADB, as well as similar triangles JLG and JCB. Finally, since

$\dfrac{GH}{GF} = \dfrac{GJ}{GE} = \dfrac{1}{2}$, we see that triangles *GJH* and *GEF* are also similar, as they share vertical angles at point *G*. Since this implies $\angle FEG = \angle HJG$ and $\angle GFE = \angle GHJ$, we have *JH* parallel to *EF*, and since *EF* is parallel to *BC*, we also have *JH*∥*BC*. We can therefore conclude that triangle *EHJ* is similar to triangle *ABC*.

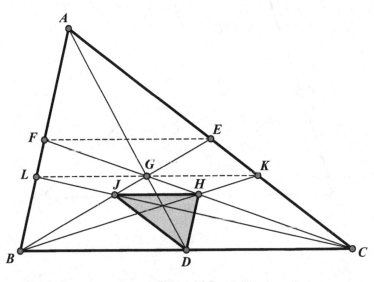

Figure 46-P

Curiosity 47. The Centroid can Provide a Most Unusual Balance

There are two parts to this proof. The first part focuses on triangle *AKG*, shown in Figure 47-P, where point *J* is the midpoint of side *AG*. We construct the perpendicular *JL* to line *MN* intersecting at point *L*. We then have $\triangle JGL \sim \triangle AGK$, which allows us to conclude that $JL = \frac{1}{2} AK$. We now focus on trapezoid *BCNM*, where *HD* is the median of the trapezoid with $HD = \frac{1}{2}(MB + NC)$. We recognize that $\triangle JLG \cong \triangle DHG$, so that $JL = DH$. Combining these equalities gives us $\frac{1}{2} AK = \frac{1}{2}(MB + NC)$ or simply $AK = MB + NC$, which was our objective.

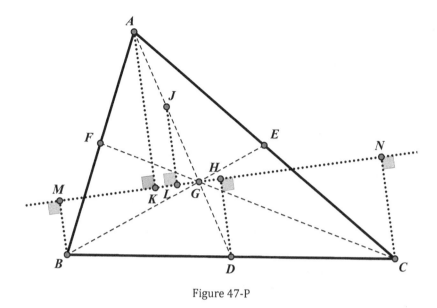

Figure 47-P

Curiosity 48. Distances from a Triangle's Vertices to a Random Line

This curiosity follows immediately from the previous one. In Figure 48-P, we have added a line $M'N'$ through point G parallel to line MKN. If AK', BM', and CN' are the perpendiculars to $M'N'$ from the vertices of triangle ABC, we know from the proof of Curiosity 47 that $AK' = M'B + N'C$. Since the lines MKN and $M'K'N'$ are parallel, we have $KK' = MM' = NN' = PG$. Since $AK' - M'B - N'C = 0$, $AK' + PG = AK' + KK' = AK$, $PG - M'B = MM' - M'B = BM$, and $PG - N'C = NN' - N'C = CN$, we have $3PG = AK' - M'B - N'C + 3PG = (AK' + PG) + (PG - M'B) + (PG - N'C) = AK + BM + CN$, which we sought to prove.

Curiosity 49. A Special Centroid Property When Two Medians are Perpendicular

Recall that the medians of a triangle trisect each other. In Figure 49-P, the relative lengths of the two parts of the medians are marked accordingly as $GE = x$, $GB = 2x$, $GD = y$ and $GA = 2y$. Applying the

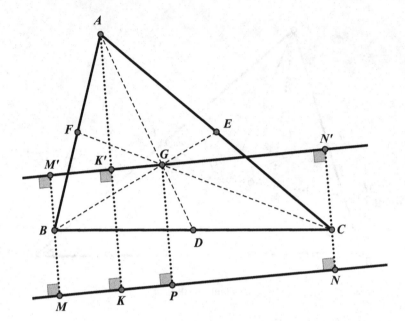

Figure 48-P

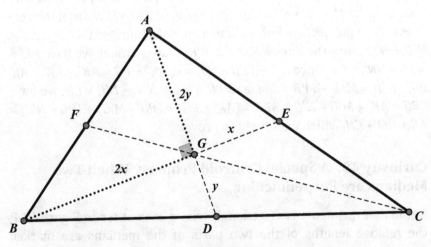

Figure 49-P

Pythagorean Theorem to triangle BGD, we get $BD^2 = y^2 + (2x)^2 = y^2 + 4x^2$.

Since D is the midpoint of BC, we have $BC^2 = (2BD)^2 = 4(y^2 + 4x^2) = 16x^2 + 4y^2$. Similarly, we can show that $AC^2 = 4x^2 + 16y^2$. Therefore, $AC^2 + BC^2 = 20(x^2 + y^2)$. Now, focusing on right triangle BAG, the Pythagorean theorem provides us with $AB^2 = (2x)^2 + (2y)^2 = 4(x^2 + y^2)$. Now multiplying by 5, we get $5AB^2 = 20(x^2 + y^2)$, and we can then conclude that $AC^2 + BC^2 = 5AB^2$.

Curiosity 50. The Centroid's Amazing Property

In order to prove that the sum of the squares of the distances from the points P and N to each of the vertices is the same, we first have to establish the following relationship, namely, that if P is any point in the plane of $\triangle ABC$ with centroid G, then $AP^2 + BP^2 + CP^2 = AG^2 + BG^2 + CG^2 + 3PG^2$.

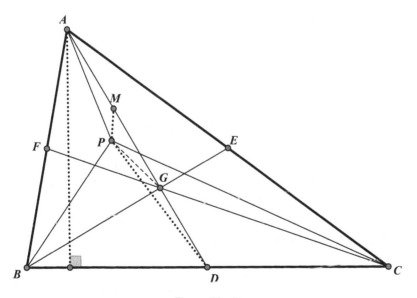

Figure 50a-P

In order to prove this, we begin by letting M be the midpoint of AG, as shown in Figure 50a-P. We first need to establish the following relationship:

Twice the square of the length of a median of a triangle equals the sum of the squares of the lengths of the two including sides minus one-half of the square of the length of the third side. Symbolically, we can write this for the median AD as $2AF^2 = AB^2 + AC^2 - \frac{1}{2}BC^2$. This can be shown by applying Stewart's theorem to $\triangle ABC$ (see Toolbox). In relation to vertex A, this theorem states that $AB^2 \cdot DC + AC^2 \cdot BD = (BD + DC)(AD^2 + BD \cdot DC)$

Let $x = DC = BD$. Then

$$xAB^2 + xAC^2 = 2x\left(AD^2 + x^2\right)$$

$$AB^2 + AC^2 = 2\left(AD^2 + x^2\right)$$

$$2AD^2 = AB^2 + AC^2 - 2x^2$$

Since $x = \frac{1}{2}BC$, we obtain our desired result $2AD^2 = AB^2 + AC^2 - \frac{1}{2}BC^2$.

We now apply this relationship to various triangles.

Applying to $\triangle PBC$, we have: $2PD^2 = PB^2 + PC^2 - \dfrac{1}{2}BC^2$. $\hspace{2em}$ (I)

Applying to $\triangle PAG$, we have: $2PM^2 = PA^2 + PG^2 - \dfrac{1}{2}AG^2$. $\hspace{2em}$ (II)

Applying to $\triangle PMD$, we have: $2PG^2 = PM^2 + PD^2 - \dfrac{1}{2}MD^2$. $\hspace{2em}$ (III)

Noting that $GD = \frac{1}{3}AD$ and $AG = \frac{2}{3}AD$ implies that $AM = MG = GD$; therefore, $MD = AG$. Substituting this in equation (III) and multiplying by 2 we get:

$$4PG^2 = 2PM^2 + 2PD^2 - AG^2 \hspace{2em} \text{(IV)}$$

Now adding (I), (II), and (IV) yields:

$$2PD^2 + 2PM^2 + 4PG^2 = BP^2 + AP^2 + 2PM^2 + CP^2 + PG^2$$

$$+2PD^2 - \frac{1}{2}BC^2 - \frac{1}{2}AG^2 - (AG)^2, \text{ or } AP^2 + BP^2 \qquad \text{(V)}$$

$$+CP^2 - 3PG^2 = \frac{3}{2}AG^2 + \frac{1}{2}BC^2.$$

A similar argument can be made for median BE, yielding:

$$AP^2 + BP^2 + CP^2 - 3PG^2 = \frac{3}{2}BG^2 + \frac{1}{2}AC^2. \qquad \text{(VI)}$$

For median CD we get:

$$AP^2 + BP^2 + CP^2 - 3PG^2 = \frac{3}{2}CG^2 + \frac{1}{2}AB^2. \qquad \text{(VII)}$$

By adding (V), (VI), and (VII), we have:

$$3\left(AP^2 + BP^2 + CP^2 - 3PG^2\right) = \frac{3}{2}\left(AG^2 + BG^2 + CG^2\right) + \frac{1}{2}\left(BC^2 + AC^2 + AB^2\right).$$

$$\text{(VIII)}$$

We now apply to $\triangle ABC$ the relationship that the sum of the squares of the lengths of the segments joining the centroid with the vertices is one-third the sum of the squares of the lengths of the sides. We can write this as $AG^2 + BG^2 + CG^2 = \frac{1}{3}\left(BC^2 + AC^2 + AB^2\right)$ or $3(AG^2 + BG^2 + CG^2) = BC^2 + AC^2 + AB^2$. This follows from the law of cosines (see Toolbox) in triangles ABD and ACD. In triangle ABD, we have $AB^2 = BD^2 + AD^2 - 2BD \cdot AD \cdot \cos\angle ADB$ and in triangle ACD, we have $AC^2 = CD^2 + AD^2 - 2CD \cdot AD \cdot \cos(180° - \angle ADB)$.

Since $CD = BD = \frac{1}{2}BC$ and $AD = \frac{3}{2}GA$, adding these two expressions gives us $AB^2 + AC^2 = \frac{1}{2}BC^2 + \frac{9}{2}GA^2$.

Similar arguments on the other two medians give us $BC^2 + AC^2 = \frac{1}{2}AB^2 + \frac{9}{2}GC^2$, and $AB^2 + BC^2 = \frac{1}{2}AC^2 + \frac{9}{2}GB^2$, and adding these three equations yields $2\left(AB^2 + BC^2 + CA^2\right) = \frac{1}{2}\left(AB^2 + BC^2 + CA^2\right) + \frac{9}{2}\left(GA^2 + GB^2 + GC^2\right)$, which simplifies to $3(AG^2 + BG^2 + CG^2) = BC^2 + AC^2 + AB^2$.

Now, we substitute this into equation (VIII) to get our desired result:

$$3\left(AP^2 + BP^2 + CP^2 - 3PG^2\right) = \frac{3}{2}\left(AG^2 + BG^2 + CG^2\right)$$
$$+ \frac{1}{2} \times 3\left(AG^2 + BG^2 + CG^2\right),$$

or $AP^2 + BP^2 + CP^2 = AG^2 + BG^2 + CG^2 + 3PG^2$.

All that now remains is to note that in Figure 50b-P, we have both $AP^2 + BP^2 + CP^2 = AG^2 + BG^2 + CG^2 + 3PG^2$ and $AN^2 + BN^2 + CN^2 = AG^2 + BG^2 + CG^2 + 3NG^2$. Since we have $PG = NG$, this allows us to conclude the desired equality: $AP^2 + BP^2 + CP^2 = AN^2 + BN^2 + CN^2$.

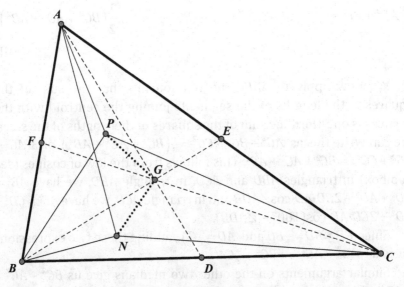

Figure 50b-P

Curiosity 51. Some Median Surprises

With the parallel lines drawn as shown in Figure 51-P, quadrilateral *AJBE* is a parallelogram. Therefore, since the diagonals of a parallelogram bisect each other, diagonal *JE* must contain the midpoint *F* of the other diagonal *AB*. This establishes that points *J*, *E*, and *F* are collinear. Since *EF* joins the midpoints of two sides of triangle *ABC*, it must be parallel to the third side, namely, *BC*, and half its length. We therefore have *JE* = *BC*, and *JBCE* is also a parallelogram since its opposite sides are equal and parallel. Since *JE* and *BC* are parallel, we have $\angle NJF = \angle CJE = \angle JCB = \angle NCD$, and with *JF* = *CD* and *JN* = *CN*, we then have $\triangle JNF \cong \triangle CND$, which gives us *DN* = *FN*.

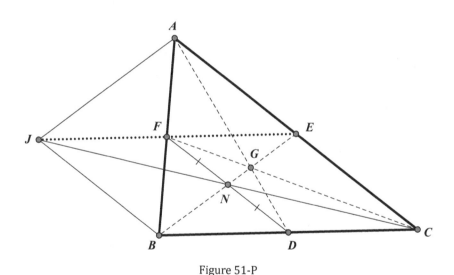

Figure 51-P

Curiosity 52. More Median Marvels

Since *H* is the common midpoint of *AC* and *BD*, we can see in Figure 52-P that what was created here is a parallelogram with $\triangle ABC \cong \triangle ACD$, which justifies having equal perimeters. We should also notice that in triangle *AFB* point *J* is the midpoint of side *AF* and point *E* is the midpoint of *FB*. Therefore, $EJ = \frac{1}{2}AB$. The same argument can be made in $\triangle AED$, where $FK = \frac{1}{2}AD = \frac{1}{2}BC$. We already know that $AH = \frac{1}{2}AC$. Therefore, we can conclude that $AB + BC + CA = AD + DC + CA = 2(AH + EJ + FK)$.

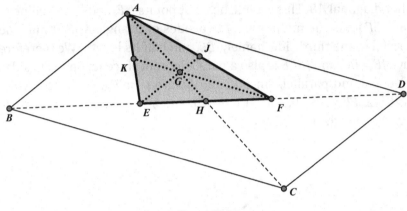

Figure 52-P

Curiosity 53. Comparing Medians to Triangle Perimeters

To show that triangle FCJ has a perimeter equal to the sum of the medians of triangle ABC, we immediately notice in Figure 53-P that CF

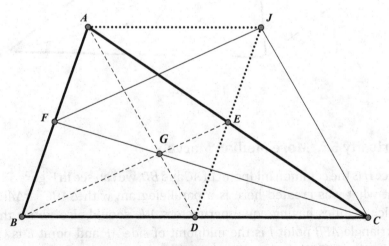

Figure 53-P

is one of the sides of triangle *FCJ*, and *CF* is also one of the medians of triangle *ABC*. In triangle *ABC*, line *DE* joins the midpoints of two sides of a triangle and is parallel to the third side *AB*, and is also one-half the length of the third side, so that *ED* = *BF*. By construction, *JE* = *ED*. Thus, *BF* = *JE*, and we can conclude that *BFEJ* is a parallelogram; therefore, *JF* = *BE*. Similarly, we can also show that *ADCJ* is a parallelogram because the diagonals *JED* and *AC* bisect each other; therefore, *JC* = *AD*. This concludes our proof that the perimeter of triangle *FCJ* is equal to the sum of the medians of triangle *ABC*.

Curiosity 54. Median Extensions Generate Collinearity

Since *AF* = *BF*, *QF* = *CF*, and ∠*AFQ* = ∠*BFC*, we can easily prove Δ*QFA* ≅ Δ*CFB*, as shown in Figure 54-P. Therefore, ∠*QAF* = ∠*CBF*, which establishes that *QA*∥*BC*. In a similar fashion, we can prove that triangles *PAE* and *BCE* are congruent, which enables us to conclude that *AP*∥*BC*. Since both *QA* and *PA* are parallel to *BC*, they must be on a common line; therefore, points *Q*, *A*, and *P* are collinear.

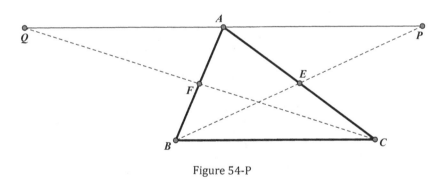

Figure 54-P

Curiosity 55. Two Unusual Triangles Share a Common Centroid

We let point *G* be the centroid of triangle *QRS*, as illustrated in Figure 55-P, and begin by recognizing that in triangle *ABC* the following proportion holds true: $\dfrac{AS}{SB} = \dfrac{BQ}{QC} = \dfrac{CR}{RA}$. Next, we locate point *P* on

BC so that $PC = BQ$. Since $\dfrac{CP}{PB} = \dfrac{BQ}{QC} = \dfrac{CR}{RA}$, this then determines that $RP\|AB$. Similarly, $\dfrac{CP}{PB} = \dfrac{BQ}{QC} = \dfrac{AS}{SB}$ also determines that $PS\|AC$. This means that $ASPR$ is a parallelogram with $AS = RP$, as well as $AS\|RP$. We now focus on triangle RQP, where we find that DK is the segment joining the midpoints of two sides of the triangle; therefore, $DK = \frac{1}{2}RP$ and $DK\|RP$. We then also have $DK\|AS$ and $DK = \frac{1}{2}AS$. We now have $\triangle GDK \sim \triangle GAS$, which enables us to conclude that SK and AD trisect each other. This means that G is the trisection point on the median AD of triangle ABC, and thus also the centroid of triangle ABC.

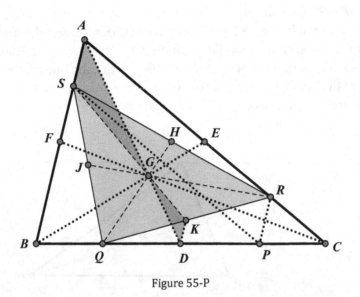

Figure 55-P

Curiosity 56. The Circumscribed Circle Revisited: The Incredible Relationship of the Centers of the Circumscribed Circles of the Median Triangles

This is quite a remarkable result, and its proof is well worth appreciating, even though it is quite long. The result itself has gained a bit of fame in recent years, and the circle containing the six circumcenters has even become decorated with the name of the person responsible for its popularization, the Dutch mathematician Floris Michel (Floor) van Lamoen (1966–), who proposed this relationship in the year 2000. It is now commonly referred to as the *Van Lamoen circle.*

As we embark on the proof of this amazing theorem, we will divide it into several parts. In a preparatory step, we first note that the three medians of a triangle can always be translated in such a way that they become the sides of a triangle themselves. This can be seen in Figure 56a-P.

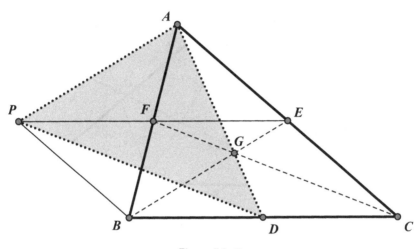

Figure 56a-P

Here, AD, BE, and CF are the medians of triangle ABC. Point P is created by translating CF to DP, or, in other words, we have $CF = DP$ and $CF \| DP$. This means that $CFPD$ is a parallelogram, and we have $CD = FP$. We know that both EF and FP are parallel to CD, and we also know that $CD = \frac{1}{2}CB = EF$; therefore, F is the midpoint of EP. Since F is the common midpoint of AB and EP, we see that $APBE$ is a parallelogram. We therefore have $PA = BE$, proving that triangle APD has sides of lengths equal to the lengths of the medians of ABC.

We are now ready for the first main part of the proof of the Van Lamoen circle.

In Figure 56b-P, we see the construction of the circumcenters, O_1, O_2, O_3, O_4, O_5, and O_6, of the six small triangles determined by the three medians of triangle *ABC*, where medians *AD*, *BE*, and *CF* meeting in the centroid *G* partition triangle *ABC* into these small triangles: *AFG*, *BGF*, *BDG*, *CGD*, *CEG*, and *AGE*.

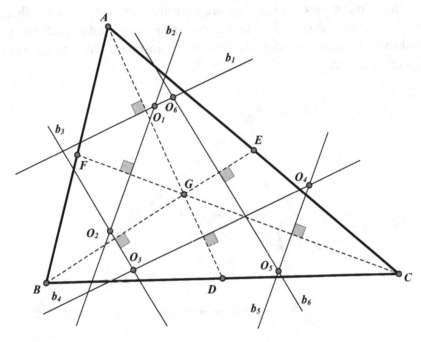

Figure 56b-P

Lines b_1, b_2, b_3, b_4, b_5, and b_6 are the bisectors of *AG*, *FG*, *BG*, *DG*, *CG*, and *EG*, respectively. This means that the common point O_1 of b_1 and b_2 is the circumcenter of triangle *AFG*, the common point O_2 of b_2 and b_3 is the circumcenter of triangle *BGF*, the common point O_3 of b_3 and b_4 is the circumcenter of triangle *BDG*, the common point O_4 of b_4 and b_5 is the circumcenter of triangle *CGD*, the common point O_5 of b_5 and b_6 is the circumcenter of triangle *CEG*, and the common point O_6 of b_6 and b_1 is the circumcenter of triangle *AGE*. The six points O_1, O_2, O_3, O_4, O_5, and O_6 are the vertices of a hexagon (not convex, in this case), whose opposite sides form parallel pairs.

We will now show that the vertices of this hexagon on parallel sides are always the vertices of an isosceles trapezoid. In Figure 56c-P, we see that points O_1, O_2, O_4, and O_5 lie on the parallel lines b_2 and b_5. We then only need to show that $O_1O_4 = O_2O_5$ holds.

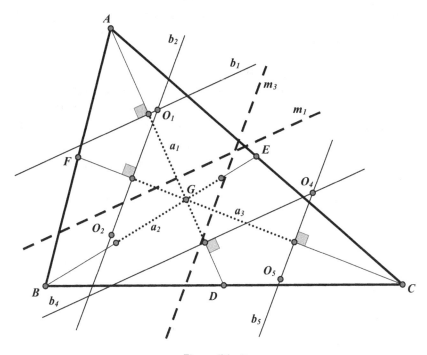

Figure 56c-P

In order to show this, we will consider the line segments a_1, a_2, and a_3 shown in Figure 56c-P. Segment a_3 connects the midpoints of CG and GF on the median CF. It is, then, perpendicular to both b_2 and b_5, and its length is half the length of CF, as b_2 is the bisector of GF, and b_5 is the bisector of CG. Similarly, a_1 connects the midpoints of AG and GD and is therefore perpendicular to b_1 and b_4 with length half that of median AD. Also, a_2 connects the midpoints of BG and GE and is thus perpendicular to b_3 and b_6 with length half that of median BE.

We have already shown that there exist triangles whose sides are equal to the lengths of the medians of ABC and whose sides are parallel to these medians. This means that there also exist triangles similar to these, whose sides are half that length, or in other words, whose sides are of lengths of a_1, a_2, and a_3. Also, in Figure 56c-P, we have added the mid-parallels m_1 and m_3 of b_1 and b_4 and b_2 and b_5, respectively.

In Figure 56d-P, we have translated a_3 parallel to m_3 in such a way that one end of the resulting segment is at point O_1, and a_1 parallel to m_1 in such a way that one end of the resulting segment also is at point O_1. The resulting triangle has sides of lengths a_1, a_2, and a_3. Since m_3 is the bisector of a_3 (and therefore also the bisector of its parallel through O_1), and m_1 is the bisector of a_1 (and of its parallel through O_1), its circumcenter is the common point X of m_3 and m_1. Furthermore, we have also created another congruent triangle. We translate a_3 parallel to m_3 in such a way that one end of the resulting segment lies in O_4, and a_1 parallel to m_1 in such a way that one end of the resulting segment also lies in O_4. the resulting triangle also has sides of lengths a_1, a_2, and a_3, and its circumcenter is the intersection point X of m_3 and m_1. The

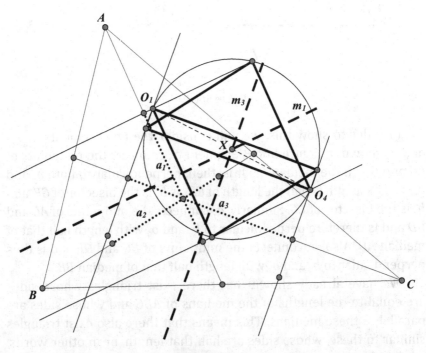

Figure 56d-P

third sides of both of these triangles have the length a_2, and they are therefore congruent.

As their circumcenters are both at X, they share the same circumcircle, whose radius is determined by the lengths a_1, a_2, and a_3 of the triangle sides. As the sides of these triangles are parallel, the triangles are symmetric with respect to X, and we see that O_1O_4 is a diameter of the circumcircle of a triangle whose sides are of length a_1, a_2, and a_3. We will let d denote the length of the diameter of such a triangle. The point X is the midpoint of O_1O_4.

Since we can recreate the identical construction with O_2 and O_5 in place of O_1 and O_4, we see that $O_1O_4 = O_2O_5 = d$, and $O_1O_2O_5O_4$ is thus indeed an isosceles trapezoid.

Furthermore, we note that the identical argument can be made for quadrilaterals $O_2O_3O_5O_6$ and $O_3O_4O_6O_1$. These are then also isosceles trapezoids with $O_2O_5 = O_3O_6 = d$ and $O_3O_6 = O_4O_1 = d$.

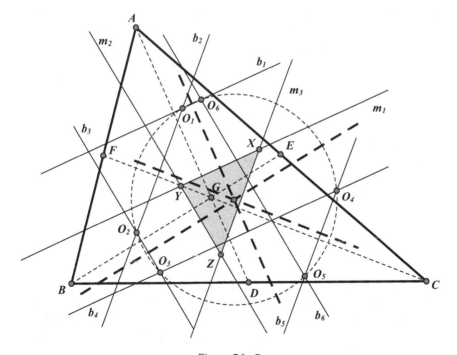

Figure 56e-P

We are now ready for the final part of our proof. In Figure 56e-P, the intersection point X of m_1 and m_3 is again the midpoint of O_1O_4. Similarly, we also see the intersection point Y of m_1 and m_2 (which is also the midpoint of O_3O_6) and the intersection point Z of m_2 and m_3 (which is also the midpoint of O_2O_5).

We can now show that the circumcenter of the triangle XYZ is equidistant from all 6 points O_1, O_2, O_3, O_4, O_5, and O_6. We have just shown that X is the midpoint of O_1O_4, and we can, similarly, show that Z is the midpoint of O_2O_5. Since quadrilateral $O_1O_2O_5O_4$ is an isosceles trapezoid, the bisector of XZ is therefore also the bisector of parallel sides O_1O_2 and O_4O_5 of the trapezoid. Since the circumcenter of triangle XYZ lies on this bisector, it is equidistant from O_1 and O_2 and also from O_4 and O_5. This argument also holds in a similar way for the other two bisectors of the sides of triangle XYZ. Since quadrilateral $O_1O_3O_4O_6$ is an isosceles trapezoid, the bisector of XY is also the bisector of parallel sides O_1O_6 and O_3O_4 of the trapezoid. Since the circumcenter of triangle XYZ lies on this bisector, it is equidistant from O_1 and O_6 and also from O_3 and O_4. Finally, since quadrilateral $O_2O_3O_5O_6$ is an isosceles trapezoid, the bisector of YZ is also the bisector of parallel sides O_2O_3 and O_5O_6 of the trapezoid. Since the circumcenter of triangle XYZ lies on this bisector, it is equidistant from O_2 and O_3 and also from O_5 and O_6.

Summing up, we see that the circumcenter of triangle XYZ is equidistant from O_1 and O_2, from O_2 and O_3, from O_3 and O_4, from O_4 and O_5, from O_5 and O_6, and from O_6 and O_1. It is thus the center of a circle containing all six points O_1, O_2, O_3, O_4, O_5, and O_6, which is what we originally set out to prove.

Curiosity 57. An Astonishing Equality

In Figure 57-P, for parallel lines EB and FC, the alternate interior angles are equal, so that $\angle BFC = \angle FBE$. We also have, in circle O, $\angle CAM = \angle EBF$, since they are both measured by the same intercepting arc ED. This means that we have $\angle DAC = \angle FBE = \angle BFC = \angle DFC$ and A, F, C, and D are concyclic, with point Q as center of the circle. Now, considering secant AM and tangent BC of circle O, we get $AM \cdot DM = BM^2$. Since

$BM = CM$, we therefore also get $AM \cdot DM = CM^2$. This implies that MC (and therefore GC) is tangent to the circle Q that contains the points A, F, C, and D.

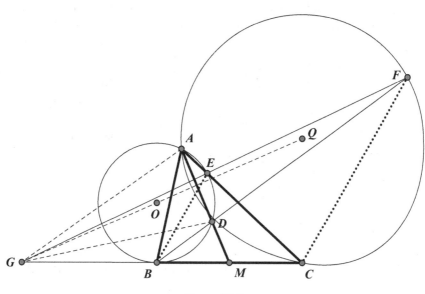

Figure 57-P

We now note that the homothety (see Toolbox) with center G that maps B onto C also maps E onto F, since BE and CF are parallel and G, E, and F lie on a common line. Since OB and QC are both perpendicular to GC, this means that the circle O is also mapped to circle Q by this homothety. Thus, we have point G on line OQ, and since OQ is the perpendicular bisector of DA, we have $GA = AD$, as we set out to prove.

Curiosity 58. An Unexpected Simultaneous Bisection and Quadrisection in a Triangle

As we see in Figure 58-P, it will be useful to draw the midpoint X of side AC, and then to rotate the triangle around X by 180°. This results in a number of parallelograms, and taking a closer look at these will provide us with the insight we need to get to prove this remarkable property.

First, we will name a few more points. We let P denote the midpoint of BN, so that we have $BP = PN = NA$ on side AB of the triangle ABC. The rotation about X maps points B, M, N, and P onto points B', M', N', and P', respectively. We then see that the rotation maps AB onto the parallel line $B'C$ and maps BC onto the parallel line AB', which means that $ABCB'$ is our first parallelogram. Since M and X are the midpoints of BC and AC, respectively, we also have $AB\|MX$, and it follows that $AB\|MX'$. Another parallelogram in our figure is, then, $ABMM'$. Since $BP = PN = NA$, we also have $B'P' = P'N' = N'C$ and $AB'\|NP'\|PN'\|BC'$. We see that the big parallelogram $ABCB'$ is, in fact, cut up into six congruent smaller parallelograms, $ANY'M'$, $M'Y'P'B'$, $NPYY'$, $Y'YN'P'$, $PBMY$, and $YMCN'$, whereby Y denotes the intersection of MM' and PN', and Y' the intersection of MM' and NP'. The lengths of the sides of each of these are $\frac{1}{3}AB$ and $\frac{1}{2}BC$. We can now discover a great deal by taking a closer look at these parallelograms.

Since the four parallelograms $NPYY'$, $Y'YN'P'$, $PBMY$, and $YMCN'$ together form a larger parallelogram $NBCP'$, its diagonals NC and BP' bisect each other. Therefore, Y is the midpoint of CN. The diagonals NY and PY' of parallelogram $NPYY'$ also bisect each other, as do the diagonals of the further composite parallelogram $ABMM'$. Because of the symmetry of the figure, the midpoints of parallelograms $ABMM'$ and

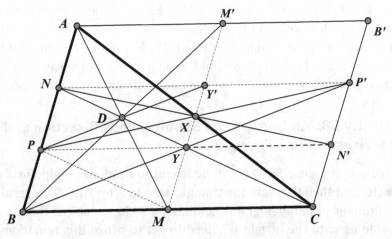

Figure 58-P

NPYY' are identical. Since this is the common midpoint of diagonal *AM* of parallelogram *ABMM'* and diagonal *NY* of parallelogram *NPYY'*, this is the point *D*. In summary, we see that point *D* is the midpoint of *AM*, and *DC* = 3*ND*, as we had set out to prove.

Curiosity 59. Noteworthy Triangle Area Relations

In Figure 59-P, we let *X* denote the intersection of *AM* and *CD*. We also have area[*DBM*] = area[*DMC*] and area[*ABM*] = area[*AMC*], since *BM* = *MC*. From this we can obtain the following equivalent equations:

$$\text{area}[ADC] - \text{area}[ABD] = 2\text{area}[ADM]$$
$$\text{area}[ADC] = 2\text{area}[ADM] + \text{area}[ABD]$$
$$\text{area}[ADX] + \text{area}[AXC] = 2\text{area}[ADX] + 2\text{area}[DMX] + \text{area}[ABD]$$
$$\text{area}[AXC] = \text{area}[ADX] + 2\text{area}[DMX] + \text{area}[ABD]$$
$$\text{area}[AMC] - \text{area}[XMC] = \text{area}[ADX] + 2 \cdot \text{area}[DMX] + \text{area}[ABD]$$

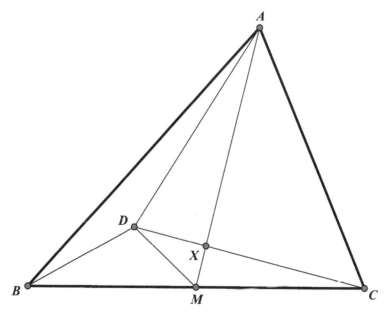

Figure 59-P

$$\text{area}[ABM] - \text{area}[XMC] = \text{area}[ADX] + 2 \cdot \text{area}[DMX] + \text{area}[ABD]$$

$$\text{area}[ABD] + \text{area}[ADX] + \text{area}[DMX] + \text{area}[DBM] - \text{area}[XMC]$$

$$= \text{area}[ADX] + 2\text{area}[DMX] + \text{area}[ABD]$$

$$\text{area}[DBM] = \text{area}[DMX] + \text{area}[XMC]$$

$$\text{area}[DBM] = \text{area}[DMC]$$

Since we know the last line to be true, we have thus established that the equivalent equality area[ADC] – area[ABD] = 2area[ADM] as we had set out to prove.

Curiosity 60. A Special Feature of a Random Point in an Equilateral Triangle

When we draw *KJ* parallel to *BC*, as in Figure 60-P, we find that it intersects altitude *AH* at point *N*; therefore, *PD = NH*. Now, considering

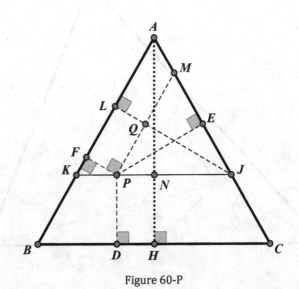

Figure 60-P

equilateral triangle *AKJ*, we draw the altitude *JL*, which intersects the line *PM*, which is parallel to *AB*, at point *Q*. Because of the resulting rectangle *FPQL*, we then have *PF* = *QL*. In equilateral triangle *MPJ*, we note that *PE* is an altitude therefore equal to the altitude *JQ*. This means that *PF* + *PE* = *QL* + *JQ* = *JL*. Finally, in the equilateral triangle *AKJ*, we have *JL* = *AN*, and thus, *PD* + *PE* + *PF* = *NH* + *AN* = *AH*.

Curiosity 61. A Special Feature of a Point Outside of an Equilateral Triangle

This can be proved with a simple application of Ptolemy's theorem (see Toolbox). Since *APCB* is a cyclic quadrilateral, as shown in Figure 61-P, we have *AP·BC* + *PC·AB* = *AC·PB*. However, since *AB* = *PC* = *AC*, we can conclude that *AP* + *PC* = *PB*.

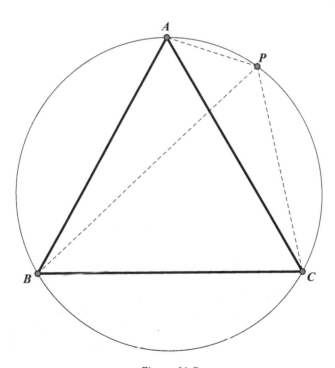

Figure 61-P

Curiosity 62. Using an Equilateral Triangle to Trisect a Line Segment

Since $DP \| AB$, we have $\angle BPD = \angle PBA = \angle DBP$, and triangle DPB is therefore isosceles with $DP = DB$, as shown in Figure 62-P. Furthermore, $\angle EDP = \angle DBP + \angle BPD = 60°$. In the same way, we also obtain $\angle DEP = 60°$, whereupon $\angle DPE = 60°$; therefore, $BD = DP = DE$. Analogously, for triangle CPE, we have $\angle EPC = \angle PCE$, and $EC = PE$. Since $DE = PE$, the result is that $BD = DE = EC$.

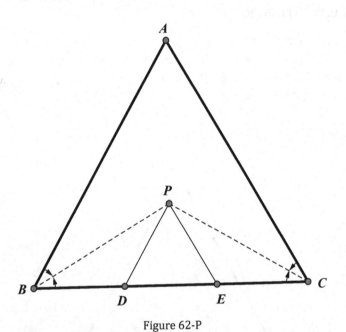

Figure 62-P

Curiosity 63. Another Way to Trisect a Line Segment with an Equilateral Triangle

Begin by noticing that $\angle QMF = 60° = \angle ACF$, and the vertical angles $\angle MFQ = \angle CFA$, as in Figure 63-P. Triangles FMQ and FCA are therefore similar. Since $AC = 2MQ$, we also have $CF = 2MF$. Put another way, we have $CF = \frac{2}{3}MC$. From this, we obtain $FC = \frac{2}{3} \cdot \frac{1}{2}BC = \frac{1}{3}BC$. Similarly, we also obtain $BE = \frac{1}{3}BC$; therefore, it must also follow that $EF = \frac{1}{3}BC$. Thus, we have shown that $BE = EF = FC$.

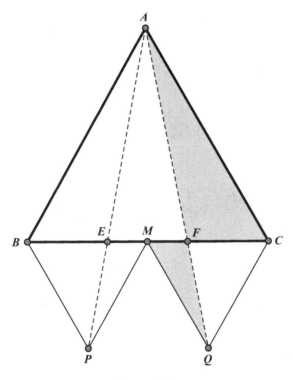

Figure 63-P

Curiosity 64. An Unusual Construction of a 30°-60°- 90° Triangle inside an Equilateral Triangle

To prove this rather unusual relationship, we will use an alternative method, which has advantages that can be seen with this proof and thus exhibits another aspect of geometry. We will use coordinate geometry, which is largely the result of the work by the famous French mathematician René Descartes (1596–1650). To begin, we place the triangles in a system of coordinates. As we see in Figure 64-P, we choose the origin at the vertex B and plot C on the x-axis. Defining $AB = BC = CA = a$ and $DB = BE = ED = b$, we then have $A\left(\dfrac{a}{2}, \dfrac{a\sqrt{3}}{2}\right)$, $B(0,0)$, $C(a,0)$, and $E(b,0)$. Since O is the center of triangle DBE, we also have $O\left(\dfrac{b}{2}, \dfrac{b\sqrt{3}}{6}\right)$ and since M is the mid-point of

AE, we have $M\left(\dfrac{a}{4}+\dfrac{b}{2}, \dfrac{a\sqrt{3}}{4}\right)$. It is now straightforward to calculate the lengths of the sides of $\triangle CMO$. We obtain

$$OM = \sqrt{\left(\frac{a}{4}\right)^2 + \left(\frac{a\sqrt{3}}{4} - \frac{b\sqrt{3}}{4}\right)^2} = \sqrt{\frac{a^2}{16} + \frac{3a^2}{16} - \frac{ab}{4} + \frac{b^2}{12}} = \frac{1}{6}\sqrt{9a^2 - 9ab + 3b^2},$$

$$OC = \sqrt{\left(a - \frac{b}{2}\right)^2 + \left(-\frac{b\sqrt{3}}{6}\right)^2} = \sqrt{a^2 - ab + \frac{b^2}{4} + \frac{b^2}{12}} = \frac{1}{3}\sqrt{9a^2 - 9ab + 3b^2} = 2 \cdot MF$$

and

$$MC = \sqrt{\left(\frac{3a}{4} - \frac{b}{2}\right)^2 + \left(-\frac{a\sqrt{3}}{4}\right)^2} = \sqrt{\frac{9a^2}{16} - \frac{3ab}{4} + \frac{b^2}{4} + \frac{3a^2}{16}}$$

$$= \frac{1}{2}\sqrt{3a^2 - 3ab + b^2} = \sqrt{3} \cdot MF.$$

As we see that $OM:OC:MC = 1:2:\sqrt{3}$, we can conclude that $\triangle CMO$ is a 30°-60°-90° triangle.

Curiosity 65. An Unusual Product of Segments in an Equilateral Triangle

We begin by drawing lines ED and FD and noticing that we will show that $\triangle BDE \sim \triangle CFD$, as shown in Figure 65-P. We know that $\angle DBE = \angle FCD = 60°$. We must now show that $\angle BED = \angle CDF$. Since EF is the perpendicular bisector of AD, we have $AE = DE$ and $AF = DF$. Therefore, triangles AEF and DFE are congruent, which tells us that $\angle FDE = \angle EAF = 60°$. Therefore,

$$\angle CDF = 180° - \angle FDE - \angle EDB = 180° - 60° - \angle EDB$$
$$= 180° - \angle DBE - \angle EDB = \angle BED,$$

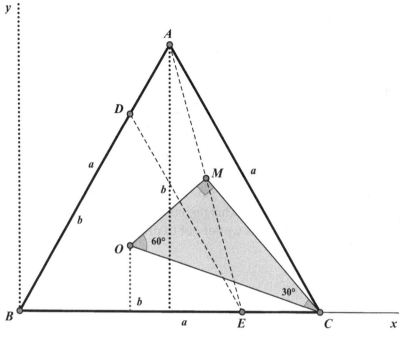

Figure 64-P

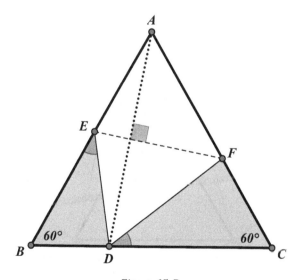

Figure 65-P

and we have $\triangle BDE \sim \triangle CFD$. This implies $\dfrac{DB}{BE} = \dfrac{FC}{DC}$, which then brings us to the sought after result, namely, that $BE \cdot FC = DB \cdot DC$.

Curiosity 66. A Surprising Relationship between an Isosceles Triangle and an Equilateral Triangle

Since *ABC* is an equilateral triangle with $AB = BC = CA$, we have $\angle BAC = 60°$. In Figure 66-P, we have added the circle with center *A* through points *B* and *C*. We are given that $\angle BAC = 2\angle BDC$, and *D* must, then, be a point on this circle, as the angle subtended by the chord *BC* is half the measure of the central angle. This means that $AD = AB$, and since $AB = BC$, we also have $AD = BC$, as we set out to prove.

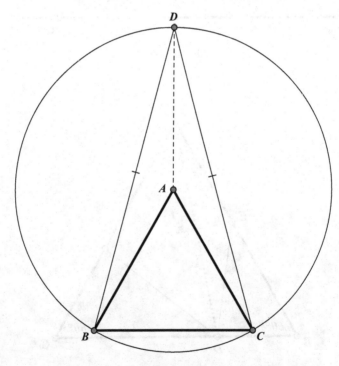

Figure 66-P

Curiosity 67. An Unexpected Equality in an Isosceles Triangle

In Figure 67-P, we have added the altitude AX of triangle ABC with X on side BC. Furthermore, Y is the point on AX where $\angle DYA = 90°$, and Z is the point on AX where $\angle EZA = 90°$. We will now show that $AE = DB$ by proving that triangles DBM and AZE are congruent.

We first note that $\angle DMB = \angle EZA = 90°$. Since AX is the altitude of isosceles triangle ABC, we have $\angle ZAE = \angle XAC = \angle BAX$, and since $AX \parallel DM$, we also have $\angle ZAE = \angle BAX = \angle BDM$. Finally, since $BM = \frac{1}{2} \cdot BP$, $NC = \frac{1}{2} \cdot PC$, and $BP + PC = BC$, we have $BM + NC = \frac{1}{2} \cdot BC$. Since $ZE + NC = XN + NC = XC = \frac{1}{2} \cdot BC$ in the isosceles triangle ABC, we have $BM + NC = ZE + NC$; therefore, $BM = ZE$. We see that triangles DBM and AZE are congruent, and this gives us that $AE = DB$.

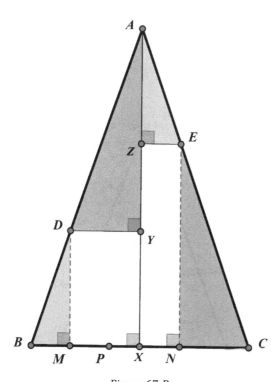

Figure 67-P

We can now show $EC = AD$ either by using an analogous argument to prove that triangles ADY and ENC are congruent or by noticing that for isosceles triangle ABC we have $AB = AC$, thus yielding $EC = AC - AE = AB - DB = AD$.

Curiosity 68. Another Unexpected Equality in an Isosceles Triangle

In Figure 68-P, we begin by drawing the perpendicular from point A to line PQ intersecting PQ at point S. We then can conclude that $AMQS$ is a rectangle, because it has right angles $\angle QMA$, $\angle SQM$, and $\angle ASQ$. This means that $AM = SQ$. We now note that triangles RSA and PAS are congruent. These triangles have a common side, AS. as well as $\angle PSA = \angle ASR = 90°$. Also, because $AM\|PQ$ and because the altitude AM bisects the angle $\angle BAC$, we have $\angle SRA = \angle MAR = \angle BAM = \angle BPS$, and we obtain $\triangle RSA \cong \triangle PAS$; thus, $PS = SR$. From this, we finally get $PQ + RQ = SQ + PS + SQ - SR = 2SQ + (PS - SR) = 2AM$.

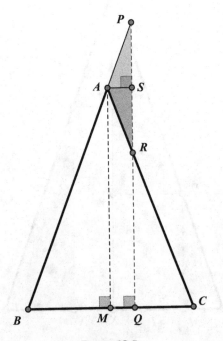

Figure 68-P

Curiosity 69. A Remarkable Geometric Equality

As shown in Figure 69-P, we first determine point G on BC (extended, if required) such that $FB = FG$. Since F is then the common midpoint of DE and BG, the quadrilateral $BEGD$ is a parallelogram with $BD = EG$ and $\angle CBA = \angle GBA = \angle BGE = \angle CGE$. Consider triangle EGC, where the vertical angles at point C yield $\angle ECG = \angle ACB$. Since triangle ABC is isosceles, we therefore have $\angle ECG = \angle ACB = \angle CBA = \angle CGE$. Triangle EGC is consequently also isosceles with $EG = CE$. Thus, we have $BD = EG = CE$, which was to be proved.

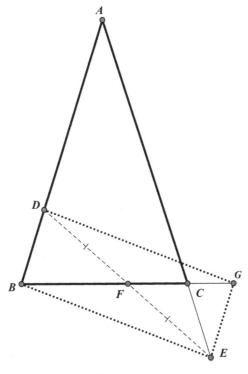

Figure 69-P

Curiosity 70. Another Remarkable Geometric Equality

To establish that $PE = BE$, we will show that triangle EBP is an isosceles right triangle. We already know $\angle BEC = \angle BEP = 90°$, since CE is the altitude to *AB*. Because ∠*ACE* = 45° in right triangle *ECA*, as illustrated in Figure 70-P, triangle *ECA* is, thus, an isosceles right triangle. Since altitude *AD* is the perpendicular bisector of *BC*, and any point on the perpendicular bisector of the line segment is equidistant from the endpoints of the line segment, we have *PB* = *PC*. Since triangle *BPC* is isosceles, we have ∠*PBD* = ∠*PCD*. When we subtract these equal angles from the base angles of the isosceles triangle *ABC*, namely, ∠*ABC* = ∠*ACB*, we get ∠*ABP* = ∠*ACP* = 45°; therefore, ∠*EPB* = 45°. Thus, we have isosceles right triangle *BEP*, so that we can conclude that *PE* = *BE*, which is what we set out to prove.

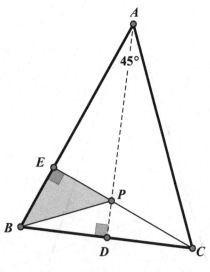

Figure 70-P

Curiosity 71. An Unexpected Perpendicularity

We begin by drawing altitudes *AM* and *BN* of triangle *ABC*, as shown in Figure 71-P. We have *MD* parallel to *BN*, so that point *D* is the midpoint of *NC*. Then ∠*CBN* + ∠*C* = 90° and ∠*MAD* + ∠*C* = 90°. Therefore, ∠*CBN* = ∠*MAD* and Δ*NBC* ~ Δ*MDA*. Since *BD* and *AF* are corresponding medians of these two similar triangles, we also have Δ*NBD* ~ Δ*FDA*. Thus, ∠*DFA* = ∠*NDB*, and so ∠*DEA* = ∠*DFA* + ∠*BDM* = ∠*NDB* + ∠*BDM* = ∠*NDM* = 90°.

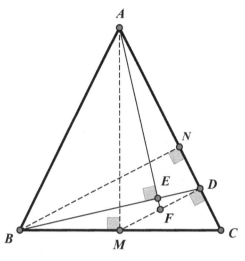

Figure 71-P

Curiosity 72. The Unexpected Property of an Altitude to the Side of an Isosceles Triangle

Considering Figure 72-P, in isosceles triangle ABC, we have $\angle ACB = \angle CBA$. However, $\angle CBA = \angle DBA + \angle CBD$, and $\angle DBA = 90° - \angle BAD$. Furthermore, we have $\angle ACB = 90° - \angle CBD$. Thus, from $\angle ACB = \angle DBA + \angle CBD$, we obtain $90° - \angle CBD = (90° - \angle BAD) + \angle CBD$. Therefore, $90° - \angle CBD = \angle DBA + \angle CBD$, or $90° - \angle DBA = 2\angle CBD$, which yields $\angle BAD = 2\angle CBD$.

Curiosity 73. Peculiar Property of Isosceles Triangles

In order to show $PD + PE = QF + QG$, we first construct BH perpendicular to AC at point H, as shown in Figure 73-P. We will show that $PD + PE = QF + QG = BH$. Choosing point P somewhere on BC, we construct JP perpendicular to BH at point J. As triangle ABC is isosceles, we have $\angle QBF = \angle CBA = \angle ACB = \angle JPB$. This means that the right triangles BPD and BPJ with the common hypotenuse BP are congruent, and we therefore have $PD = BJ$. We also note that quadrilateral $JPEH$ is

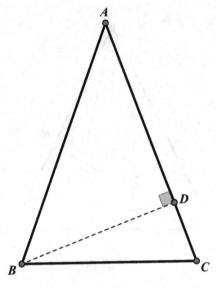

Figure 72-P

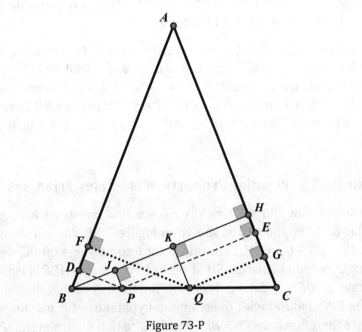

Figure 73-P

a rectangle; therefore, *PE* = *JH*. Thus, we have *PD* + *PE* = *BJ* + *JH* = *BH*. Since this argument was independent of the choice of the position of point *P* on *BC*, we can also choose point *Q* on *BC* and point *K* on *BH* with *QK* ⊥ *BH* to show *QF* + *QG* = *BH* in a completely analogous way. Therefore, *PD* + *PE* = *QF* + *QG* for any two randomly chosen points *P* and *Q* along the base *BC* of isosceles triangle *ABC*.

Curiosity 74. A Bizarre Connection: The Triangle Incenter on its Circumcircle

In Figure 74-P, we assume that *I* is the point in which the circumcircle of triangle *DBC* intersects the altitude of *ABC* at point *A*. We wish to show that this point is then the incenter of triangle *ADC*. In order to demonstrate this, we have added line segments *IB*, *IC*, and *ID*. Since two angles measured by the same arc BC are equal, we have ∠*BIC* = ∠*BDC* = 180° − ∠*CDA*. Since point *I* lies on the altitude of the

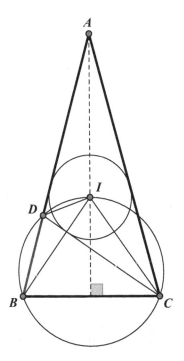

Figure 74-P

isosceles triangle *ABC*, triangle *IBC* is also isosceles, and we have $\angle CBA = \frac{1}{2} \cdot (180° - \angle BIC) = \frac{1}{2} \cdot \angle CDA$. Now, since *B*, *C*, *I*, and *D* lie on a common circle, we therefore have $\angle CDI = \angle CBI = \frac{1}{2} \cdot \angle CDA$, and point *I* then lies on the angle bisector of $\angle CAD$. Since *I* also lies on the altitude of isosceles triangle *ABC* from point *A*, which is also the angle bisector of $\angle DAC$, point *I* is the intersection of two angle bisectors of triangle *ADC* and thus the incenter of triangle *ADC*.

Curiosity 75. Collinear Points Generate an Angle Bisector

We notice immediately that quadrilateral *AFPE* is cyclic in Figure 75-P, since one pair of opposite angles are right angles. Therefore, since inscribed angles measured by the same arc are equal, we have $\angle PAE = \angle PFE$, and similarly, $\angle FAP = \angle FEP$. If we can show that $\angle PFE = \angle FEP$, then we will have shown that $\angle FAP = \angle PAE$ and *AP*, therefore, bisects $\angle BAC$. The line *RQS* is drawn parallel to *BC* and is therefore bisected by the median *AM*. Since $QD \perp RS$, triangle *RPS* is isosceles with $PR = PS$ and $\angle PRS = \angle RSP$. Noting that quadrilaterals *PQRF* and *PSEQ* are also cyclic with pairs of opposite right angles, we have $\angle PRS = \angle PFE$ and $\angle RSP = \angle FEP$. This means that we have established that $\angle FAP = \angle PAE$; therefore, *AP* bisects $\angle BAC$.

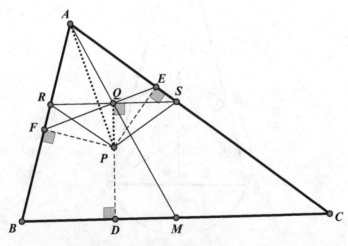

Figure 75-P

Curiosity 76. Unexpected Similar Triangles

We first note that the two triangles *ABM* and *ACM* in Figure 76-P share an angle at point *M*. Thus, to prove similarity, we merely need to show that these two triangles have another pair of equal angles. Since *AH* is the external angle bisector of triangle *ABC* in vertex *A*, we have $\angle BAH = 90° + \frac{1}{2} \cdot \angle BAC$. In triangle *BHA*, we obtain the following:

$$\angle AHB = 180° - \angle HBA - \angle BAH = 180° - \angle CBA - \left(90° + \frac{1}{2} \cdot \angle BAC\right)$$

$$= 90° - \angle CBA - \frac{1}{2} \cdot \angle BAC.$$

The angle bisectors *AG* and *AH* at vertex *A* are perpendicular since they are bisecting adjacent supplementary angles. Point *M* is the midpoint of *GH*, so it follows that point *A* must lie on the circle with center *M* and diameter *GH*. We have *MA = MH* as they are both radii of this circle, making triangle *MHA* isosceles, so that $\angle MAH = \angle AHM = 90° - \angle CBA - \frac{1}{2} \cdot \angle BAC$.

This gives us $\angle CAM = \angle BAH - \angle BAC - \angle MAH = \left(90° + \frac{1}{2} \cdot \angle BAC\right) - \angle BAC - \left(90° - \angle CBA - \frac{1}{2} \cdot \angle BAC\right) = \angle CBA$, and consequently, triangles *ABM* and *ACM* are similar.

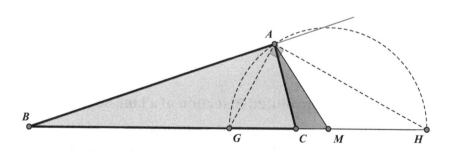

Figure 76-P

Curiosity 77. Perpendiculars to Four Angle Bisectors of a Triangle Reveal Four Collinear Points

In Figure 77-P, consider △*ABC*, where *BD* and *BE* are the interior and exterior angle bisectors (respectively) at vertex *B*, along with *D* and

E being the feet of the perpendiculars to *BD* and *BE* from vertex *A*. Extend *AD* and *AE* to meet *BC* at *M* and *N*, respectively. Since *BD* is the bisector of ∠*B* and *AM* ⊥ *BD*, we know that Δ*BDA* ≅ Δ*BMD*, and so we then have *AD* = *DM*. Similarly, since *BE* is the (external) angle bisector of ∠*B* and *AN* ⊥ *BE*, we also know that Δ*BAE* ≅ Δ*BEN*; therefore, *AE* = *EN*. We then have *DE* as the midline of Δ*AMN*. We know that the midline of a triangle is parallel to the third side (in this case, *DE*∥*NBMC*) and bisects the other two sides of the triangle, or can be considered to be halfway between the vertex and the opposite side of the triangle.

Analogously, the feet of the perpendiculars from *A* to the interior and exterior bisectors of ∠*C* are also on a line that is parallel to *NBMC* and lies halfway between *BC* and *A*. Therefore, all four feet of the perpendiculars from vertex *A* to the interior and exterior bisectors of the other two angles of Δ*ABC* are collinear.

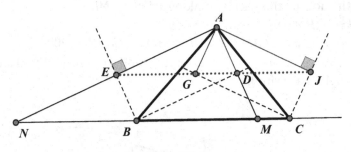

Figure 77-P

Curiosity 78. A Convoluted Bisection of a Line Segment

In order to simplify notation a bit, we define the lengths of the sides of triangle *ABC* as *BC* = *a*, *CA* = *b*, and *AB* = *c*, as shown in Figure 78-P.

Since *CD* and *CE* are the angle bisectors of triangle *ABC* in *C*, we can apply the angle bisector properties (see Toolbox) and obtain $\frac{BD}{AD} = \frac{BC}{AC} = \frac{a}{b} = \frac{BE}{AE}$. Furthermore, since *DQ* is parallel to *BC*, we note that triangles *ADQ* and *ABC* are similar, as we also have similar triangles *APB* and *AQD*. From Δ*ADQ* ~ Δ*ABC* we obtain $\frac{BC}{DQ} = \frac{AB}{AD}$. Also, from $\triangle APB \sim \triangle AQD$ we obtain $\dfrac{BP}{DQ} = \dfrac{BE}{BD}$. Having established this, we will now express each of the fractions $\dfrac{BC}{DQ}$ and $\dfrac{BP}{DQ}$ in terms of a, b, and c, which will allow us to compare their relative sizes more clearly.

First, we note that $\dfrac{BD}{AD} = \dfrac{a}{b}$, and $BD + AD = AB = c$, which implies $\dfrac{c - AD}{AD} = \dfrac{a}{b}$ or $bc - b{\cdot}AD = a{\cdot}AD$, and which gives us $AD = \dfrac{bc}{a+b}$.

From this, we obtain $\dfrac{BC}{DQ} = \dfrac{AB}{AD} = \dfrac{c}{\dfrac{bc}{a+b}} = \dfrac{a+b}{b}$.

Similarly, $\dfrac{BE}{AE} = \dfrac{a}{b}$, and $AE - BE = AB = c$, which implies $\dfrac{BE}{BE + c} = \dfrac{a}{b}$ or $b(BE) = a(BE) + ac$, and which gives us $BE = \dfrac{ac}{b-a}$.

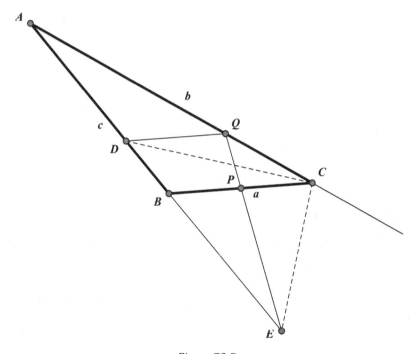

Figure 78-P

We also have $DB = AB - AD = c - \dfrac{bc}{a+b} = \dfrac{ac}{a+b}$ and $DE = BE + BD =$

$\dfrac{ac}{b-a} + \dfrac{ac}{a+b} = \dfrac{2abc}{(a+b)(b-a)}$, and from this we obtain $\dfrac{BP}{DQ} = \dfrac{BE}{DE} =$

$\dfrac{\dfrac{ac}{b-a}}{\dfrac{2abc}{(a+b)(b-a)}} = \dfrac{a+b}{2b}$.

We see that $\dfrac{BC}{DQ} = \dfrac{a+b}{b} = 2\dfrac{a+b}{2b} = 2\dfrac{BP}{DQ}$. This implies that $BC = 2BP$. We have thus shown that P is the midpoint of BC, which is what we had to prove.

Curiosity 79. Determining the Perimeter of a Triangle Without Measuring its Side Lengths

We begin this proof by drawing perpendicular lines GF and GK to lines AD and BC, respectively, as well as drawing line AG in Figure 79-P. Since right triangles CKG and CGH share the side CG and $\angle KCG = \angle GCH$, we have $\triangle CKG \cong \triangle CGH$; therefore, $CH = CK$, and $GH = GK$. Analogously, we find that $\triangle BFG \cong \triangle BGK$, so that $BK = BF$, and $GK = GF$. Therefore, $GH = GF$. Furthermore, the two right triangles that share the common side AG enable us to have $\triangle AFG \cong \triangle AGH$, and subsequently $AF = AH$. Also, $AH = AC + CH = AC + CK$. Analogously, $AF = AB + BK$. Summarizing, we have $AH = \frac{1}{2}(AH + AF) = \frac{1}{2}(AC + CK + BK + AB) = \frac{1}{2}(AC + BC + AB)$, which is what we set out to prove.

Curiosity 80. A Strange Angle Trisection

Noting that in Figure 80-P point R is the midpoint of AB and that AB is twice as long as BC, we have $AR = BR = BC$. Furthermore, we are given $\angle PRA = \angle BRP = \angle PCB = 90°$. We now take a closer look at triangles PAR, PRB, and PBC. Recognizing that PR is a common side of triangles PAR and PRB, we have $\triangle PAR \cong \triangle PRB$. Also, seeing that PB is a common side of triangles PRB and PBC, we also have $\triangle PRB \cong \triangle PBC$. Since the three triangles PAR, PRB, and PBC are congruent, we obtain

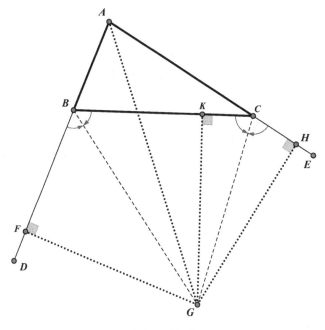

Figure 79-P

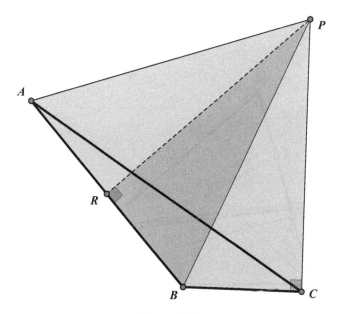

Figure 80-P

∠APR = ∠RPB = ∠BPC, and we have shown that PR and PB trisect the angle ∠APC.

Curiosity 81. A Most Unexpected Equality

To simplify the notation, we let ∠A, ∠B, and ∠C denote the interior angles of △ABC, and to further simplify our discussion, we will use ∠P = ∠BPC and ∠Q = ∠BQC. As shown in Figure 81-P, in triangle ABX, we have ∠QBA = ∠Q − ∠A, and since BX is the angle bisector, this gives us $\angle XBA = \frac{1}{2}(\angle Q - \angle A)$. In triangle ABX, we then have $\angle AXB = 180° - \angle A - \angle XBA = 180° - \angle A - \frac{1}{2}(\angle Q - \angle A)$. Repeating this on the other side gives us $\angle CYA = 180° - \angle A - \frac{1}{2}(\angle P - \angle A)$, and this allows us to do the following calculation in quadrilateral AYRX:

$$\angle BRC = \angle XRY$$
$$= 360° - \angle A - \angle AXR - \angle RYA$$
$$= 360° - \angle A - \angle AXB - \angle CYA$$
$$= 360° - \angle A - \left(180° - \angle A - \frac{1}{2}(\angle Q - \angle A)\right) - \left(180° - \angle A - \frac{1}{2}(\angle P - \angle A)\right)$$
$$= \frac{1}{2}(\angle P + \angle Q).$$

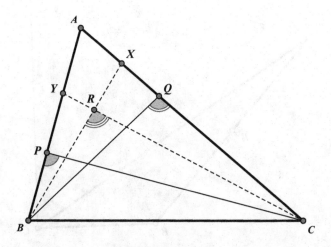

Figure 81-P

We have thus obtained $\frac{1}{2}\left(\angle BPC + \angle BQC\right) = \angle BRC$, which can also be expressed as $\angle BPC + \angle BQC = 2\angle BRC$.

Curiosity 82. A Triangle Peculiarity

Since points F and D are the midpoints of sides AB and BC, respectively, as shown in Figure 82-P, we find that DF is parallel to AC and $DF = \frac{1}{2}AC = CE$. This allows us to conclude that quadrilateral $CEFD$ is a parallelogram; therefore, $\angle DFE = \angle ECD$. In right triangle AHC, the line segment EH is the median to the hypotenuse AC; therefore, $EH = EC$, and $\angle ECH = \angle CHE$. Now considering right triangle CHB, where DH is the median to hypotenuse BC, we have $\angle HCD = \angle DHC$. By addition, we get $\angle ECD = \angle ECH + \angle HCD = \angle CHE + \angle DHC = \angle DHE$. However, we know from parallelogram $CEDF$ that $\angle ECF = \angle EDF$, so we can conclude that $\angle DFE = \angle DHE$.

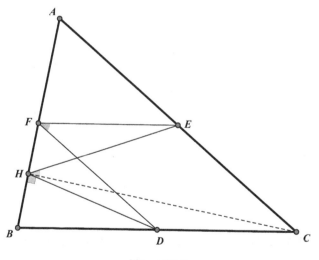

Figure 82-P

Curiosity 83. A Counterintuitive Area Equality of Triangles

In order to simplify notation a bit, in Figure 83-P, we denote $BC = a$, $CA = b$, $AB = c$, and $AD = FC = x$. Since $AB \| DE \| FG$, we have three similar triangles *ABC*, *DEC*, and *FGC*. We therefore have $\dfrac{DE}{AB} = \dfrac{DC}{AC} = \dfrac{b-x}{b}$ or

$DE = \dfrac{c(b-x)}{b}$, and $\dfrac{FG}{AB} = \dfrac{FC}{AC} = \dfrac{x}{b}$ or $FG = \dfrac{cx}{b}$.

Since *DE∥FG*, we have ∠*ADE* = ∠*AFG*. We now merely have to calculate the areas. We have area$[AED] = \dfrac{1}{2}(AD)(DE)\sin\angle ADE =$

$\dfrac{1}{2}x\dfrac{c(b-x)}{b}\sin\angle ADE = \dfrac{1}{2}\dfrac{cbx-cx^2}{b}\sin\angle ADE$ and also area$[AGF] =$

$\dfrac{1}{2}(AF)(FG)\sin\angle AFG = \dfrac{1}{2}(b-x)\dfrac{cx}{b}\sin\angle AFG = \dfrac{1}{2}\dfrac{cbx-cx^2}{b}\sin\angle AFG$. From the above equalities, since we earlier showed that ∠*ADE* = ∠*AFG*, we can then conclude that area[*AED*] = area[*AGF*].

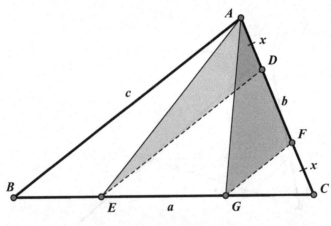

Figure 83-P

Curiosity 84. Parallel Lines Create a Double Area Triangle

In Figure 84-P, point *E* lies on the segment *AC*, while points *D* and *F* are outside the triangle *ABC*. (For any other configuration, the proof can be adapted to conform to the labeling.) We extend *EB* to intersect *DF* in point *P*. Since *AD∥EP∥CF*, we have $\dfrac{PF}{DF} = \dfrac{EC}{AC}$. We have *EB* is parallel to *AD*, so that triangles *CEB* and *CAD* are similar, whereupon

$\dfrac{EB}{AD} = \dfrac{EC}{AC}$. Similarly, since BP is parallel to AD, triangles FBP and FAD are similar, and it follows that $\dfrac{BP}{AD} = \dfrac{PF}{DF}$. From the previous three proportions we have $\dfrac{EB}{AD} = \dfrac{EC}{AC} = \dfrac{PF}{DF} = \dfrac{BP}{AD}$, from which it follows that $EB = BP$. We now only need to note that triangles ABE, DBE, and DPB all have bases of equal length, which are either EB or BP, and equal altitudes, since $AD \| EP$. This gives us area$[ABE]$ = area$[DBE]$ = area$[DPB]$. Also, triangles CEB, FEB, and FBP all have bases of equal length, namely, either EB or BP, and equal altitudes, since $CF \| EP$. This gives us area$[CEB]$ = area$[FEB]$ = area$[FBP]$. Summarizing, we have

$$\text{area}\left[DFE\right] = \text{area}\left[DBE\right] + \text{area}\left[DPB\right] + \text{area}\left[FBP\right] + \text{area}\left[FEB\right]$$

$$= 2\text{area}\left[ABE\right] + 2\text{area}\left[CEB\right] = 2\left(\text{area}\left[ABE\right] + \text{area}\left[CEB\right]\right) = 2\text{area}\left[ABC\right]$$

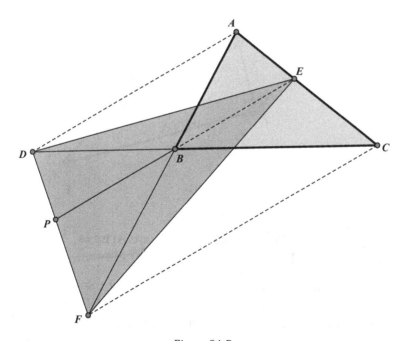

Figure 84-P

and we see that the area of triangle *DFE* is twice the area of triangle *ABC*.

Curiosity 85. Unexpected Triangle Area Relationships

If we wish to prove area[*BDC*] + area[*ADC*] = area[*ABD*], we can note that this is equivalent to area[*BDC*] + area[*ADC*] = area[*ABC*] + area[*BCD*] – area[*ADC*] or 2area[*ADC*] = area[*ABC*]. This follows when we consider the altitude *DM* of equilateral triangle *BDC*, as shown in Figure 85-P. Since *BDC* is equilateral, *M* is the midpoint of side *BC*, and we have *BC* = 2*MC*. Since *DM*∥*CA*, triangles *ADC* and *AMC* have the same base *AC*, the same altitude, and therefore the same area. We thus obtain area[*ABC*] = 2area[*AMC*] = 2area[*ADC*], and we have shown area[*BDC*] + area[*ADC*] = area[*ABD*].

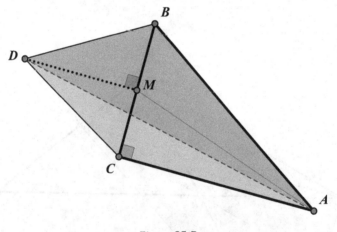

Figure 85-P

Curiosity 86. Creating a Triangle Whose Area is Three-Quarters the Area of a Given Triangle

In Figure 86-P, we want to prove that the sides of triangle *GFC* have lengths equal to the medians of triangle *ABC*. Side *FC* is already a median of *ABC*, which covers one side of the triangle. Since *D* and *E* are midpoints of *BC* and *CA*, respectively, we then have *DE*∥*AB*. Since

FG was constructed parallel to *BE*, we see that *BEGF* is a parallelogram, and the length of *FG* is therefore equal to median *BE*. Finally, we see that *EF*∥*BD* also holds true for the same reason, and this implies *DE* = *BF* = *EG*. Since *E* is the midpoint of *AC*, quadrilateral *ADCG* is also a parallelogram, and we then have *GC* = *AD*. Thus, the sides of triangle *GFC* are equal to the medians of *ABC*.

Now, we shall consider the area property. Since *ED*∥*AB*, triangles *CED* and *CAB* are similar, and the median *CP* of triangle *CED* lies on the median *CF* of triangle *CAB*. It follows that *P* is the midpoint of *DE*, and since *DE* is half as long as *AB*, we see that *EP* is half as long as *AF*. We can now compare the areas of triangles *AFC* and *EFC*. These triangles have a common side *CF*, and we have ∠*CPE* = ∠*CFA* and *AF* = 2*EP*. The altitude of triangle *AFC* is therefore twice that of triangle *EFC*, and from this we obtain area$\left[EFC\right]=\frac{1}{2}\cdot$area$\left[AFC\right]$. Now, we can also compare the areas of triangles *GFC* and *EFC*. We have already established *DE* = *EG* and *DP* = *PE*. This means that we also have *GP* = 3*EP*, and since triangles *GFC* and *EFC* have the common side *FC*, this implies area$\left[EFC\right]=\frac{1}{3}\cdot$area$\left[GFC\right]$. Also, we know that the median *CF* of triangle *ABC* divides it into triangles of equal area, and, we then have area$\left[AFC\right]=\frac{1}{2}\cdot$area$\left[ABC\right]$. Summarizing, we have area$\left[GFC\right]=3$area$\left[EFC\right]=\frac{3}{2}$area$\left[AFC\right]=\frac{3}{4}$area$\left[ABC\right]$, which we had set out to prove.

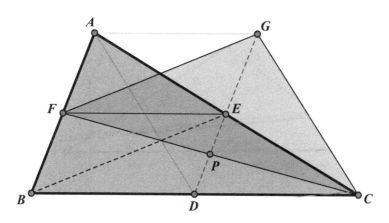

Figure 86-P

Curiosity 87. An Astounding Construction: Similar Triangles Whose Area Ratio is 1:4

As shown in Figure 87-P, point J is the foot of the perpendicular from P onto AB extended, so that we have $\angle AJP = 90°$, and since point H is the foot of the perpendicular from P onto AC, we have $\angle PHA = 90°$. Segment AP is therefore a diameter of the circumcircle of $PJAH$, and point K, which is the center of the circle, is then the midpoint of AP. Analogous arguments for the other two circles show us that L is the midpoint of BP and N the midpoint of CP. This means that triangle KLN is obtained from triangle ABC by homothety (see Toolbox) with center P and factor $\frac{1}{2}$. From this fact, we can deduce that triangles ABC and KLN are similar, and that the sides of KLN are half as long as those of ABC. The ratio of the areas of two similar triangles is equal to the square of their

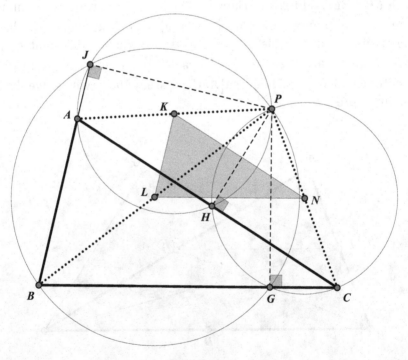

Figure 87-P

corresponding sides. Consequently, area$[KLN] = \frac{1}{4} \cdot$ area$[ABC]$, which completes the proof.

Curiosity 88. Unforeseen Equality of Inscribed Triangles

In order to simplify notation, we set $a = BC$, $b = CA$, $c = AB$, $x = BR = CD$, $y = CS = AE$, and $z = AT = BF$, as shown in Figure 88-P. We wish to prove area[DEF] = area[RST]. Since

$$\text{area}[DEF] = \text{area}[ABC] - \text{area}[AFE] - \text{area}[BDF] - \text{area}[CED]$$

and

$$\text{area}[RST] = \text{area}[ABC] - \text{area}[ATS] - \text{area}[BRT] - \text{area}[CSR],$$

this is equivalent to showing that

$$\text{area}[AFE] + \text{area}[BDF] + \text{area}[CED] = \text{area}[ATS] \\ + \text{area}[BRT] + \text{area}[CSR].$$

We can write

$$\text{area}\big[AFE\big] + \text{area}\big[BDF\big] + \text{area}\big[CED\big]$$
$$= \frac{1}{2}y(c-z)\sin \angle A + \frac{1}{2}z(a-x)\sin \angle B + \frac{1}{2}x(b-y)\sin \angle C$$

and

$$\text{area}\big[ATS\big] + \text{area}\big[BRT\big] + \text{area}\big[CSR\big]$$
$$= \frac{1}{2}z(b-y)\sin \angle A + \frac{1}{2}x(c-z)\sin \angle B + \frac{1}{2}y(a-x)\sin \angle C.$$

The law of sines (see Toolbox) in $\triangle ABC$ gives us $a \cdot \sin \angle B = b \cdot \sin \angle A$, $b \cdot \sin \angle C = c \cdot \sin \angle B$ and $c \cdot \sin \angle A = a \cdot \sin \angle C$, and we can use this to show that

$$\frac{1}{2}y(c-z)\sin\angle A+\frac{1}{2}z(a-x)\sin\angle B+\frac{1}{2}x(b-y)\sin\angle C$$

$$=\frac{1}{2}\left(yc\sin\angle A-yz\sin\angle A+za\sin\angle B-zx\sin\angle B+xb\sin\angle C-xy\sin\angle C\right)$$

$$=\frac{1}{2}\left(ya\sin\angle C-yz\sin\angle A+zb\sin\angle A-zx\sin\angle B+xc\sin\angle B-xy\sin\angle C\right)$$

$$=\frac{1}{2}\left(zb\sin\angle A-zy\sin\angle A+xc\sin\angle B-xz\sin\angle B+ya\sin\angle C-yx\sin\angle C\right)$$

$$=\frac{1}{2}z(b-y)\sin\angle A+\frac{1}{2}x(c-z)\sin\angle B+\frac{1}{2}y(a-x)\sin\angle C.$$

We have thus shown that

$$\text{area}[AFE] + \text{area}[BDF] + \text{area}[CED] = \text{area}[ATS]$$
$$+ \text{area}[BRT] + \text{area}[CSR],$$

which enables us to conclude that area[DEF] = area[RST].

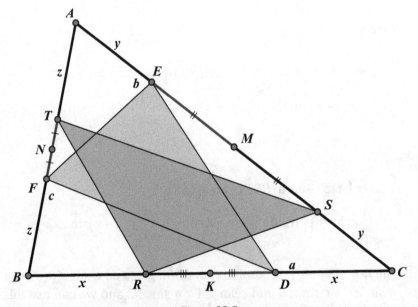

Figure 88-P

Curiosity 89. The Unanticipated Commonality of Equal Area Triangles

In order to prove the bisection of *AD* at point *M*, we first need a few auxiliary lines. In Figure 89-P, we construct a point *F* with *BF* parallel to *AC* and *CF* parallel to *AB*, which results in parallelogram *ABFC*, and then we have triangles *ABC* and *FBC* with equal areas. We are given that triangles *ABC* and *BDC* also have equal areas. It then follows that the areas of triangles *BDC* and *BFC* are also equal. That allows us to conclude that *FD* must be parallel to *BC*, since these triangles share a common base *BC* and the altitudes of these two equal-area triangles must also be equal. Since the diagonals of a parallelogram bisect each other in parallelogram *ABFC*, point *G* is then the midpoint of diagonal *AF*. *F* In triangle *ADF*, line segment *GM* is parallel to the base *FD*; therefore, point *M* must be the midpoint of *AD*, which is what we set out to prove.

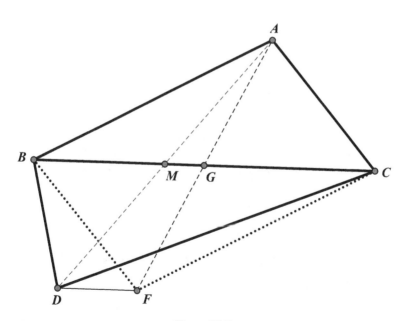

Figure 89-P

Curiosity 90. How a Random Point Divides the Area of a Triangle in Half

Since $MB = \frac{1}{2}AB$ as shown in Figure 90-P, and since $\triangle CAB$ and $\triangle CMB$ share a common altitude CG relative to their common base on line AB, we can conclude that $\text{area}[CMB] = \frac{1}{2}\text{area}[CAB]$. Furthermore, because $MN \parallel PC$, we notice that $\triangle PMN$ and $\triangle CMN$ have the same altitude to their common base MN, and are therefore also equal in area. We now notice that area$[CMB]$ = area$[NMB]$ + area$[CMN]$. Thus, by substitution, we get

$$\text{area}[PBN] = \text{area}[NMB] + \text{area}[PMN] = \text{area}[NMB] + \text{area}[CMN]$$

$$= \text{area}[CMB] = \frac{1}{2} \cdot \text{area}[CAB]$$

which is what we set out to prove.

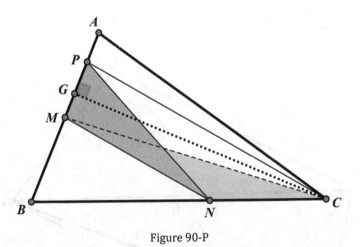

Figure 90-P

Curiosity 91. Determining a Triangle One-Third of the Area of a Given Triangle

In Figure 91-P, the altitude AQ of triangle ABC from vertex A has been added, as has the altitude ER of triangle BGE from vertex E.

Furthermore, we have point P as the intersection of AQ and EH. Noting that the homothety with center A and factor $\frac{2}{3}$ (see Toolbox) maps BC to EH, we see that $BC \| EH$. $ERQP$ is therefore a rectangle. Also, the homothety maps Q onto P. We can now calculate the following:

$$\text{area}[BGE] = \frac{1}{2} BG \cdot ER = \frac{1}{2}\left(\frac{2}{3}BC\right) \cdot PQ = \frac{1}{2}\left(\frac{2}{3}BC\right) \cdot \left(\frac{1}{3}AQ\right)$$

$$= \frac{2}{9}\left(\frac{1}{2}BC \cdot AQ\right) = \frac{2}{9} \cdot \text{area}[ABC].$$

Analogously, we can also draw GD and calculate $\text{area}[CJG] = \frac{2}{9} \cdot \text{area}[ABC]$, or draw JF and calculate $\text{area}[AEJ] = \frac{2}{9} \cdot \text{area}[ABC]$. We then see that area[$AEJ$] = area[$BGE$] = area[$CJG$]. Furthermore, we now have

$$\text{area}[EGJ] = \text{area}[ABC] - \text{area}[EBG] - \text{area}[CJG] - \text{area}[AEJ]$$

$$= \text{area}[ABC] - \frac{2}{9}\text{area}[ABC] - \frac{2}{9}\text{area}[ABC] - \frac{2}{9}\text{area}[ABC]$$

$$= \frac{1}{3}\text{area}[ABC].$$

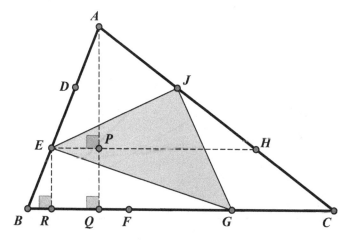

Figure 91-P

Curiosity 92. Trisection Points Partitioning a Triangle

Since points D and E cut off proportional segments along the sides AB and AC of triangle ABC, DE is parallel to BC. It follows that triangles ADE and ABC are similar, and we have $\dfrac{DE}{BC} = \dfrac{AD}{AB} = \dfrac{1}{3}$. We also have $\triangle EDF \sim \triangle CBF$, because of the equal alternate interior angles formed by the parallel lines DE and BC, with their corresponding sides in the ratio 1:3. Consequently, we have $\dfrac{DF}{FC} = \dfrac{1}{3}$. Since triangles BFD and BCF share the same altitude BP from B to DC, as shown in Figure 92-P, their areas are also in the ratio 1:3.

To simplify matters, we will let area[BFD] = R. Then area[BCF] = $3R$, and area[BCD] = $4R$.

Triangles BCD and BCA share the same altitude CQ from C to AB; therefore, the ratio of their areas is equal to $BD:BA$ = 2:3. We find that twice the area of $\triangle ABC$ equals three times the area of $\triangle BCD$, which equals $12R$, and from this we get area[ABC] = $6R$. Since area[BCF] = $3R$, we then have area$\left[BCF\right] = \frac{1}{2} \cdot$area$\left[ABC\right]$. Applying the same argument to the other side of the triangle, $\triangle ABE$ and $\triangle ABC$

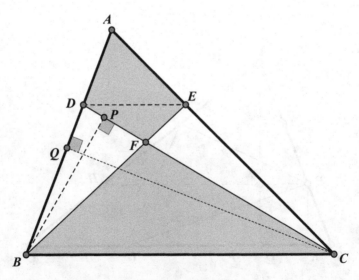

Figure 92-P

share the same altitude from B to side CA, and we therefore have area[ABE]:area[ABC] = AE:AC = 1:3. Using our relative areas, we then have 3·area[ABE] = area[ABC] = $6R$, or area[ABE] = $2R$. This gives us area[$ADFE$] = area[ABE] – area[BFD] = $2R$ – R = R = area[BFD]. In an analogous fashion, we can also show area[$ADFE$] = area[CEF], and this gives us area[$ADFE$] = area[BFD] = area[CEF].

Curiosity 93. Further Surprises Provided by the Trisection Points of Triangle Sides

In Figure 93a-P, we have drawn line segments AK, WL, HM, and DN perpendicular to BC, which we will use to calculate the area of triangle WBC to determine its fraction of the area of triangle ABC. If we consider altitude WL of triangle WBC as a multiple of AK, which is the altitude of triangle ABC, we have $WL = a{\cdot}AK$. This enables us to get: $\text{area}\left[WBC\right]=\frac{1}{2}BC \cdot WL =\frac{1}{2}BC \cdot a \cdot AK =a\left(\frac{1}{2}BC \cdot AK\right)=a \cdot \text{area}\left[ABC\right]$, and we then only need to express WL as a multiple of AK. To achieve this goal, we first consider line segment HM. Since triangles CAK and CHM are similar right triangles with $CH=\frac{1}{3}{\cdot}CA$, we also have $HM=\frac{1}{3} \cdot AK$. Similarly, since triangles BKA and BND are similar right triangles with $DB=\frac{2}{3} \cdot AB$, we also have $DN=\frac{2}{3} \cdot AK$. Furthermore, it will be useful to note that similar triangles CAK and CHM also give us $CM=\frac{1}{3} \cdot CK$, and the similar triangles BKA and BND also give us $BN=\frac{2}{3} \cdot BK$. Next, we turn our attention to triangles WBL and HBM. These are also similar right triangles, and we obtain $\dfrac{BL}{BM}=\dfrac{WL}{HM}$ or

$$BL=\frac{WL}{HM} \cdot BM =\frac{a \cdot AK}{\frac{1}{3}AK}\left(BK +CK -CM\right)=3a\left(BK +\tfrac{2}{3}CK\right).$$

In an analogous way, similar right triangles WLC and DNC give us

$$\frac{CL}{CN}=\frac{WL}{DN} \quad \text{or} \quad CL=\frac{WL}{DN} \cdot CN =\frac{a \cdot AK}{\frac{2}{3}AK}\left(BK +CK -BN\right)=\tfrac{3}{2}a\left(\tfrac{1}{3}BK +CK\right).$$

Since $BL + CL = BK + CK$, this gives us $3a\left(BK +\frac{2}{3}CK\right)+\frac{3}{2}a\left(\frac{1}{3}BK +CK\right)=BK +CK$. This is equivalent to $\frac{7}{2}a\left(BK +CK\right)=BK +CK$, and then we have $a=\frac{2}{7}$. Recalling area[WBC] = a·area[ABC], we have thus found that the $\text{area}\left[WBC\right]=\frac{2}{7} \cdot \text{area}\left[ABC\right]$. Analogously, we can also show that the $\text{area}\left[UCA\right]=\frac{2}{7} \cdot \text{area}\left[VAB\right]$, and that the $\text{area}\left[WBC\right]=\frac{2}{7} \cdot \text{area}\left[ABC\right]$.

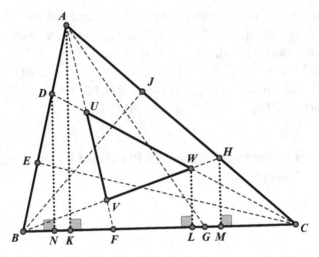

Figure 93a-P

As seen in Figure 93b-P, we can consequently calculate the area of triangle UVW with respect to triangle ABC since area$[UVW]$ = area$[ABC]$ – area$[WBC]$ – area$[UCA]$ – area$[VAB]$. We then obtain area$[UVW]$ = area$[ABC] - \frac{2}{7}$area$[ABC] - \frac{2}{7}$area$[ABC] - \frac{2}{7}$area$[ABC] = \frac{1}{7}$area$[ABC]$.

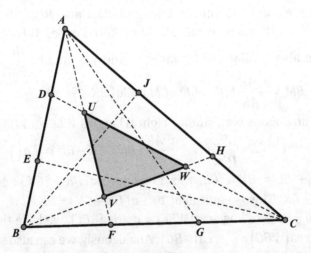

Figure 93b-P

Similarly, in Figure 93c-P, we can also calculate the area of triangle *RST* as follows:

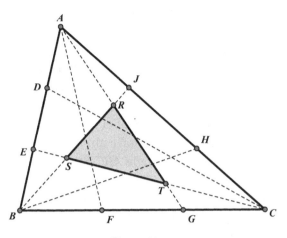

Figure 93c-P

Analogous calculations give us area$[SBC]$ = area$[TCA]$ = area$[RAB]$ = $\frac{2}{7}$·area$[ABC]$, and we therefore obtain area$[RST]$ = area[*ABC*] − area[*SBC*] − area[*TCA*] − area[*RAB*]. Thus, area$[RST]$ = area$[ABC]$ − $\frac{2}{7}$·area$[ABC]$ − $\frac{2}{7}$·area$[ABC]$ − $\frac{2}{7}$·area$[ABC]$ = $\frac{1}{7}$·area$[ABC]$.

We see that the areas of both triangles *UVW* and *RST* are equal to one-seventh the area of triangle *ABC*, as we set out to prove.

Curiosity 94. Another Surprise Provided by the Trisection Points of Triangle Sides

The first part of this proof is quite similar to the proof of Curiosity 93. In Figure 94-P, point *P* is the intersection of lines *BJ* and *CD*, and we have line segments *AK, XL, JM,* and *PN* perpendicular to *BC*. We wish to express *XL* as a multiple of the altitude *AK* of triangle *ABC* as *XL* = *a·AK*. As in the proof of Curiosity 94, we note that $JM = \frac{2}{3} \cdot AK$ and $CM = \frac{2}{3} \cdot CK$. Considering the similar right triangles *XBL* and *JBM*, we obtain

$$\frac{BL}{BM} = \frac{XL}{JM} \quad \text{or} \quad BL = \frac{XL}{JM}BM = \frac{a \cdot AK}{\frac{2}{3}AK}\left(BK + CK - CM\right) = \frac{3}{2}a\left(BK + \frac{1}{3}CK\right).$$

Furthermore, the similar right triangles XLF and AKF give us $\dfrac{FL}{FK} = \dfrac{XL}{AK}$

or $FL = \dfrac{XL}{AK} \cdot FK = \dfrac{a \cdot AK}{AK} (FB - BK) = a\left(\frac{1}{3}(BK + CK) - BK\right) = a\left(\frac{1}{3}CK - \frac{2}{3}BK\right)$.

Since $BL + FL = \frac{1}{3} \cdot (BK + CK)$, we get $\frac{3}{2}a\left(BK + \frac{1}{3}CK\right) + a\left(\frac{1}{3}CK - \frac{2}{3}BK\right) = \frac{1}{3}(BK + CK)$. This is equivalent to $\frac{5}{6}a(BK + CK) = \frac{1}{3}(BK + CK)$, or $a = \frac{2}{5}$. We have thus shown that $XL = \frac{2}{5} \cdot AK$. Since we can repeat this argument in an analogous way for point Z, replacing BJ with CD and AF with AG, we see that the perpendicular distance from Z to BC is also equal to $\frac{2}{5}AK$. Line segment XZ is therefore parallel to line segment BC, with the distance between the two being equal to $\frac{2}{5}AK$.

In the proof to Curiosity 92, we have already shown that the area of triangle PBC is half that of triangle ABC. Since these two triangles share the base BC, the altitude of PBC is equal to $\frac{1}{2}AK$. Noting now that triangles PBC and PXZ are similar, as $BC \| XZ$, and the altitude of triangle PXZ is equal to $\frac{1}{2}AK - \frac{2}{5}AK = \frac{1}{10}AK$, we see that the altitude of PBC is five times the altitude of triangle PXZ, as $5 \cdot \frac{1}{10} \cdot AK = \frac{1}{2} \cdot AK$. This means that we also have $XZ = \frac{1}{5} \cdot BC$.

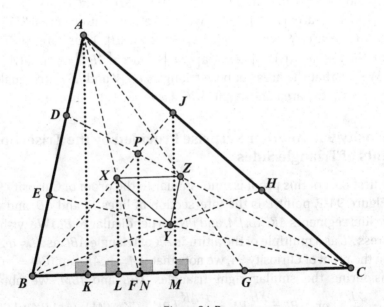

Figure 94-P

This argument can be repeated in analogously for the other sides of triangle XYZ, and we also obtain $XY \| CA$ with $XY = \frac{1}{5} \cdot CA$ and $YZ \| AB$ with $YZ = \frac{1}{5} \cdot AB$. We see that triangles ABC and XYZ are similar, and the lengths of the sides of triangle XYZ are one-fifth of the lengths of the sides of the corresponding sides of triangle ABC. The ratio of the areas of the similar triangles is the square of the ratio of the corresponding side lengths, and we thus obtain $\text{area}[XYZ] = \left(\frac{1}{5}\right)^2 \cdot \text{area}[ABC] = \frac{1}{25} \cdot \text{area}[ABC]$, which completes the proof.

Curiosity 95. Yet Another Surprise Provided by the Trisection Points of Triangle Sides

This proof is nearly identical to the proof of Curiosity 94, although a bit shorter. In Figure 95-P, we have added the line segments AK, MX, and EY perpendicular to BC. We express MX as a multiple of the altitude AK of triangle ABC as $MX = a \cdot AK$. From the similar triangles EBY and ABK we have $EY = \frac{1}{3} AK$ and $BY = \frac{1}{3} BK$. Considering the similar right triangles MXC and EYC, we obtain $\dfrac{CX}{CY} = \dfrac{MX}{EY}$ or

$$CX = \frac{MX}{EY} \cdot CY = \frac{a \cdot AK}{\frac{1}{3} AK}\left(CK + BK - BY\right) = 3a\left(\tfrac{2}{3} BK + CK\right).$$

The similar right triangles MXF and AKF give us $\dfrac{FX}{FK} = \dfrac{MX}{AK}$ or

$$FX = \frac{MX}{AK} \cdot FK = \frac{a \cdot AK}{AK}\left(FB - BK\right) = a\left(\tfrac{1}{3}\left(BK + CK\right) - BK\right) = a\left(\tfrac{1}{3} CK - \tfrac{2}{3} BK\right).$$

Since $CX - FX = CF = \frac{2}{3}\left(BK + CK\right)$, we get $3a\left(\tfrac{2}{3} BK + CK\right) - a\left(\tfrac{1}{3} CK - \tfrac{2}{3} BK\right) = \frac{2}{3}\left(BK + CK\right)$. This is equivalent to $\frac{8}{3} a\left(BK + CK\right) = \frac{2}{3}\left(BK + CK\right)$, or $a = \frac{1}{4}$. We have thus shown that $MX = \frac{1}{4} \cdot AK$. Since we can repeat this argument in an analogous way for point N, replacing CE with BH, we see that the perpendicular distance from N to BC is also equal to $\frac{1}{4} \cdot AK$. Line segment MN is therefore parallel to line segment BC, with the distance between the two being equal to $\frac{1}{4} \cdot AK$.

Noting now that triangles AFG and AMN are similar, as $FG \| MN$, and the altitude of AMN is equal to $AK - \frac{1}{4} \cdot AK = \frac{3}{4} \cdot AK$, we see that the altitude of AMN is $\frac{1}{3}$ times that of AFG. This means that we also have $MN = \frac{3}{4} FG = \frac{3}{4}\left(\frac{1}{3} BC\right) = \frac{1}{4} BC$.

This argument can be repeated in analogously for the other sides of triangle *LMN*, and we then obtain *NL*∥*CA* with $NL = \frac{1}{4} \cdot CA$ and *LM*∥*AB* with $LM = \frac{1}{4} \cdot AB$. We see that triangles *ABC* and *LMN* are similar, and the lengths of the sides of triangle *XYZ* are one-quarter of the lengths of the corresponding sides of triangle *ABC*. The ratio of the area of the similar triangles is known to be the square of the ratio of their corresponding side lengths, and we thus obtain $\text{area}[LMN] = \left(\frac{1}{4}\right)^2 \cdot \text{area}[ABC] = \frac{1}{16} \cdot \text{area}[ABC]$, which completes the proof.

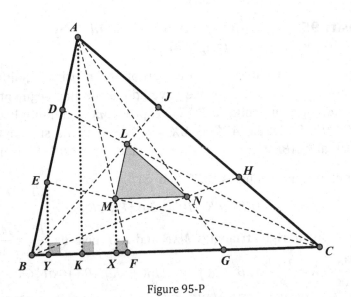

Figure 95-P

Curiosity 96. Medians and Trisectors Partitioning a Triangle with an Unexpected Result

As this proof follows immediately from the results of Curiosities 94 and 95, it is perhaps one of the most astonishing results of all. As we see in Figure 96-P, we can calculate the area of the small triangle *LUZ* as area[*LUZ*] = area[*ABC*] – area[*UAB*] – area[*ZBC*] – area[*LCA*]. Since *AM* and *BN* are medians of triangle *ABC*, their intersection, *U*, is the centroid of triangle *ABC*. As we know from Curiosity 42, the areas of *UAB*, *UBC*, and *UCA* are equal, and we therefore have area$[UAB] = \frac{1}{3} \cdot$ area$[ABC]$. In Curiosity 94, we saw that the perpendicular distance of Z from BC is $\frac{2}{5}$ that of the distance from A to BC, which enables us to get area$[ZBC] = \frac{2}{5} \cdot$ area$[ABC]$. Finally, in Curiosity 95, we saw that the perpendicular distance of L from CA is $\frac{1}{4}$ that of the distance from B to CA, and this enables us to get area$[LCA] = \frac{1}{4} \cdot$ area$[ABC]$. In summary, we therefore have

$$\text{area}[LUZ] = \text{area}[ABC] - \text{area}[UAB] - \text{area}[ZBC] - \text{area}[LCA]$$

$$= \text{area}[ABC] - \frac{1}{3} \cdot \text{area}[ABC] - \frac{2}{5} \cdot \text{area}[ABC] - \frac{1}{4} \cdot \text{area}[ABC]$$

$$= \frac{1}{60} \cdot \text{area}[ABC].$$

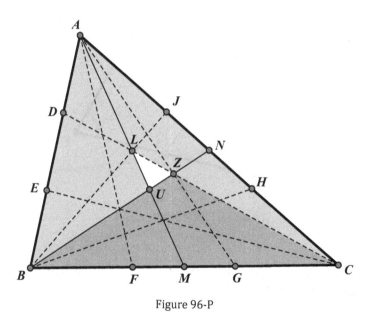

Figure 96-P

Curiosity 97. The Unforeseen Characteristic of a Random Triangle with a 60° Angle

To justify this astonishing result created by the angle bisectors BP and CQ of angles B and C, respectively, which intersect in point E, we need to first construct a few auxiliary lines. Selecting point R on side BC so that $BR = BQ$, as we see in Figure 97-P, we draw lines PQ, QR, RP, and ER. Since $\angle A = 60°$, we know that $\angle B + \angle C = 120°$. Therefore, because of the angle bisectors, $\angle CBE + \angle ECB = 60°$. Thus, $\angle BEC = 120°$ and $\angle CEP = \angle QEB = 60°$. Since we chose R such that $BR = BQ$ and BE bisects $\angle B$, we have $\triangle BRE \cong \triangle BEQ$. Therefore, $EQ = ER$ and $\angle QEB = \angle BER = 60°$. Thus, $\angle REC = \angle BEC - \angle BER = 120° - 60° = 60° = \angle CEP$. Also, CE bisects $\angle C$, which allows us to conclude that $\triangle CER \cong \triangle CPE$; thus, $CP = CR$. Therefore, $BC = BR + CR = BQ + CP$.

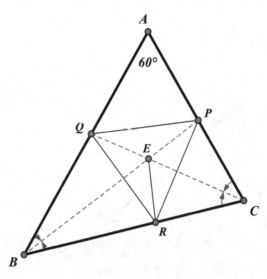

Figure 97-P

Curiosity 98. Another Surprising Feature of a Random Triangle with a 60° Angle

In Figure 98-P, quadrilateral $AFHE$ is cyclic since the opposite angles are right angles. Since $\angle BAC = 60°$, we obtain $\angle EHF = 180° - \angle BAC = 120°$. That makes $\angle FHB = 60°$, so that our objective is to show that $\angle OHB = 30°$. Since $\angle BAC = 60°$, we have $\angle BOC = 120°$. However, $\angle BHC$ is supplementary to $\angle FHB$ and therefore also equal to $120°$. Because we have $\angle BOC = \angle BHC$, quadrilateral $BCHO$ is cyclic, and from this we also obtain $\angle OHB = \angle OCB$. Since triangle OBC is isosceles with $\angle BOC = 120°$, its base angles are equal to $30°$, and we, then, obtain $\angle OHB = \angle OCB = 30°$; thus, $\angle FHO = \angle FHB - \angle OHB = 60° - 30° = 30°$, which implies that OH bisects $\angle FHB$.

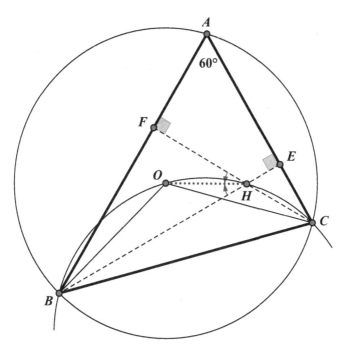

Figure 98-P

Curiosity 99. Another Unexpected Collinearity

In Figure 99-P, we have added line segments CI and CD. Since BE is the angle bisector of $\angle CBA$, and points D and E both lie on the circumcircle of ABC, we have $\angle CDE = \angle CBE = \angle EBA = \angle EDA$. Thus, we have DE as the angle bisector of $\angle CDA$. We consider triangle DCI. We have $\angle CDI = \angle CDA = \angle CBA = \angle B$ and $\angle ICD = \angle ICB + \angle BCD = \angle ICB + \angle BAD = \frac{1}{2}\angle C + \frac{1}{2}\angle A$. Therefore, we have $\angle DIC = 180° - \angle CDI - \angle ICD = \angle A + \angle B + \angle C - \angle B - \left(\frac{1}{2}\angle C + \frac{1}{2}\angle A\right) = \frac{1}{2}\angle C + \frac{1}{2}\angle A$, and we see that triangle DCI is isosceles. The angle bisector DE of $\angle CDA$ is therefore the perpendicular bisector of the base *CI*. Next, we consider the angles of triangle *PIC*, recalling that *PI* was defined as parallel to *AB*, and *PC* as the tangent of the circumcircle of *ABC* at point *C*. Since *PC* is the tangent of the circle *O*, we have $\angle PCI = \angle PCB - \angle ICB = 180° - \angle A - \frac{1}{2} \cdot \angle C$. Consider the intersection of *IP* and *AC* at point *Q*. We have $\angle IQC = \angle BAC = \angle A$; thus, in triangle *QIC*, we obtain $\angle CIP = \angle CIQ = 180° - \angle IQC - \angle QCI = 180° - \angle A - \frac{1}{2} \cdot \angle C$. We see that triangle *PIC* is also isosceles, which means that point *P* also lies on the perpendicular bisector of base *IC*. We see that points *D*, *E*, and *P* lie on a common line, as we had set out to prove.

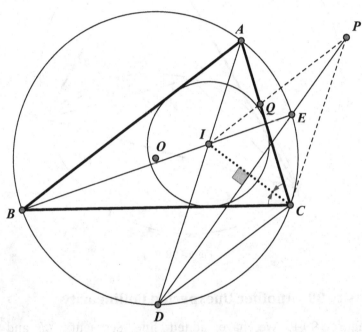

Figure 99-P

Curiosity 100. Reflections of Triangles Generate Concurrent Circles and Concurrent Lines

In Figure 100-P, we define point *P* as the intersection of the circumcircles of *AC'B* and *ACB'*. Our goal is to show that this point *P* also lies on the circumcircle of *A'CB*.

Since the opposite angles of a cyclic quadrilateral are supplementary, we have $\angle APB = 180° - \angle BC'A = 180° - \angle C$, and $\angle CPA = 180° - \angle AB'C = 180° - \angle B$. We then have $\angle BPC = 360° - \angle APB - \angle CPA = 360° - (180° - \angle C) - (180° - \angle B) = \angle B + \angle C = 180° - \angle A$.

Since $\angle CA'B = \angle A$, this implies that P must lie on the circumcircle of $A'CB$, completing the first part of the proof.

It now remains to be shown that lines AQ, BR, and CS are concurrent. To do this, we will apply Ceva's theorem (see Toolbox), and we let X, Y, and Z denote the intersections of AQ and BC, BR and CA, and CS and AB, respectively, as shown in Figure 100-P. We then need to show that $\dfrac{BX}{CX} \cdot \dfrac{CY}{AY} \cdot \dfrac{AZ}{BZ} = 1$. In triangle ABX, the law of sines (see Toolbox) gives us $\dfrac{BX}{AX} = \dfrac{\sin \angle BAX}{\sin \angle B}$, and, similarly, in triangle AXC we get $\dfrac{CX}{AX} = \dfrac{\sin \angle XAC}{\sin \angle C}$. Taking the quotient of these two equations yields $\dfrac{BX}{CX} = \dfrac{\sin \angle C \cdot \sin \angle BAX}{\sin \angle B \cdot \sin \angle XAC}$. Since Q is the center of the circumcircle of $A'CB$, we have $\angle CQB = 2\angle CA'B = 2\angle B$. Triangle QCB is isosceles; therefore, we get $\angle QBC = \frac{1}{2} \cdot (180° - \angle CQB) = \frac{1}{2} \cdot (180° - 2 \cdot \angle A) = 90° - \angle A$. From this, we obtain $\angle QBA = \angle QBC + \angle CBA = 90° - \angle A + \angle B$, and applying the law of sines to triangle ABQ yields $\dfrac{BQ}{AQ} = \dfrac{\sin \angle BAX}{\sin \angle QBA} = \dfrac{\sin \angle BAX}{\sin(90° - \angle A + \angle B)}$. Analogously, the same argument for triangle AQC yields $\dfrac{CQ}{AQ} = \dfrac{\sin \angle XAC}{\sin(90° - \angle A + \angle C)}$. Since $BQ = CQ$, these two expressions are equal, and we obtain $\dfrac{\sin \angle BAX}{\sin(90° - \angle A + \angle B)} = \dfrac{\sin \angle XAC}{\sin(90° - \angle A + \angle C)}$, which is equivalent to $\dfrac{\sin \angle BAX}{\sin \angle XAC} = \dfrac{\sin(90° - \angle A + \angle B)}{\sin(90° - \angle A + \angle C)}$. Substitution gives us $\dfrac{BX}{CX} = \dfrac{\sin \angle C \cdot \sin \angle BAX}{\sin \angle B \cdot \sin \angle XAC} = \dfrac{\sin \angle C \cdot \sin(90° - \angle A + \angle B)}{\sin \angle B \cdot \sin(90° - \angle A + \angle C)}$.

If we repeat the argument we have just made for AQ in an analogous way for lines BR and CS, we obtain $\dfrac{CY}{AY} =$

$$\dfrac{\sin\angle A \cdot \sin\left(90° - \angle B + \angle C\right)}{\sin\angle C \cdot \sin\left(90° - \angle B + \angle A\right)} \quad \text{and} \quad \dfrac{AZ}{BZ} = \dfrac{\sin\angle B \cdot \sin\left(90° - \angle C + \angle A\right)}{\sin\angle A \cdot \sin\left(90° - \angle C + \angle B\right)}.$$

Multiplication then gives us the following:

$$\dfrac{BX}{CX} \cdot \dfrac{CY}{AY} \cdot \dfrac{AZ}{BZ} =$$

$$\dfrac{\sin\angle C \cdot \sin\left(90° - \angle A + \angle B\right)}{\sin\angle B \cdot \sin\left(90° - \angle A + \angle C\right)} \cdot \dfrac{\sin\angle A \cdot \sin\left(90° - \angle B + \angle C\right)}{\sin\angle C \cdot \sin\left(90° - \angle B + \angle A\right)} \cdot \dfrac{\sin\angle B \cdot \sin\left(90° - \angle C + \angle A\right)}{\sin\angle A \cdot \sin\left(90° - \angle C + \angle B\right)} =$$

$$\dfrac{\sin\angle C \cdot \sin\left(90° - \angle A + \angle B\right)}{\sin\angle B \cdot \sin\left(90° - \left(\angle A - \angle C\right)\right)} \cdot \dfrac{\sin\angle A \cdot \sin\left(90° - \angle B + \angle C\right)}{\sin\angle C \cdot \sin\left(90° - \left(\angle B - \angle A\right)\right)} \cdot \dfrac{\sin\angle B \cdot \sin\left(90° - \angle C + \angle A\right)}{\sin\angle A \cdot \sin\left(90° - \left(\angle C - \angle B\right)\right)}$$

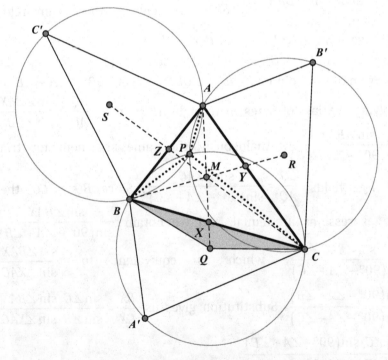

Figure 100-P

We know that $\sin(90° + x) = \sin\pi(90° - x)$; thus, we see that all of these expressions cancel, and the product is equal to 1. By Ceva's theorem, lines AX, BY, and CZ are therefore concurrent, and this also is true for AQ, BR, and CS, as we had set out to prove.

Curiosity 101. The Wonders of Three Concurrent Congruent Circles

To begin we let r denote the length of the common radius of the three congruent circles. As we see in Figure 101a-P, point P lies on all three circles, and we have $PQ = PR = PS$. Point P is therefore the center of the circumcircle of triangle QRS, and the radius of this circle is equal to r. This completes the proof to part a).

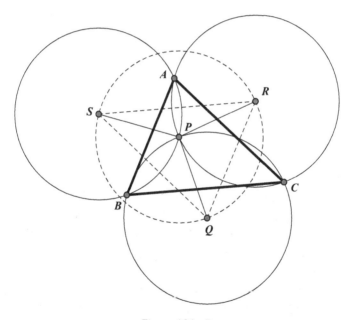

Figure 101a-P

The next step will be a bit more complicated. To prove part b), we consider the parts of the configuration shown in Figure 101b-P, where we have chosen A', B', and C' such that $A'P$, $B'P$, and $C'P$ are diameters of the circles with centers Q, R, and S, respectively. Also, we have chosen point D as the intersection of AP extended with BC, and point E as the intersection of BP extended with CA, and further point F as the intersection of CP extended with AB.

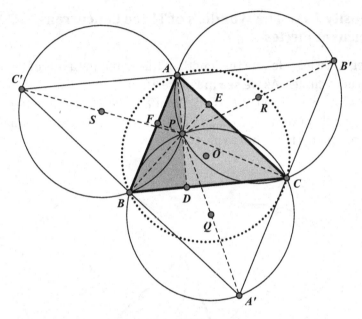

Figure 101b-P

Since PB' and PC' are diameters of congruent circles, we have $PB' = PC'$. Therefore, triangle $PB'C'$ is isosceles with $\angle C'B'P = \angle PC'B'$. Because of the equal angles subtended over AP in the circles with centers R and S, we then obtain $\angle PBA = \angle PC'A = \angle PC'B' = \angle C'B'P = \angle AB'P = \angle ACP$, and we can name this angle $\angle PBA = \angle ACP = x$. Analogously, we also obtain $\angle PCB = \angle BAP = y$ and $\angle PAC = \angle CBP = z$.

We now consider the shaded triangles ABD and ADC. In ABD we have $180° = \angle ADB + \angle BAD + \angle DBA = \angle ADB + \angle BAP + \angle CBP + \angle PBA = \angle ADB + y + z + x$.

Similarly, in ADC we have

$$180° = \angle CDA + \angle ACD + \angle DAC = \angle CDA + \angle ACP$$
$$+ \angle PCB + \angle PAC = \angle CDA + x + y + z.$$

This implies that $\angle ADB = \angle CDA$, and since these two angles are supplementary, they must each equal 90°; thus, $AP \perp BC$. Analogously, we can also show $BP \perp CA$ and $CP \perp AB$. Thus, have proved that point P is the orthocenter of triangle ABC, completing the proof of part b).

Now that we have established that P is the orthocenter of ABC, we can consider part c). In Figure 101c-P, we see triangle ABC along with the circle with center Q and point P, the orthocenter of triangle ABC. Furthermore, we have added the intersection of AD extended with the circumcircle of ABC, which we have labeled as P^*.

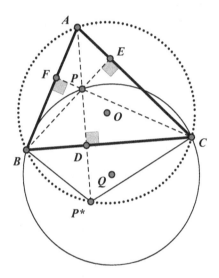

Figure 101c-P

We will now show that triangles PBC and P^*CB are symmetric with respect to their common side BC. Since PP^* is perpendicular to BC, we must show that $\angle BPC = \angle CP^*B$, as the symmetry then follows. Since quadrilateral ABP^*C is cyclic, the opposite angles are supplementary, so that $\angle CP^*B = 180° - \angle A$. From the right triangles BCE and BCF we get:

$$\angle BPC = 180° - \angle CBP - \angle PCB = 180° - (90° - \angle C)$$
$$- (90° - \angle B) = \angle B + \angle C = 180° - \angle A,$$

and we have shown that $\angle BPC = \angle CP^*B$. Since this implies the congruence of triangles PBC and P^*CB, their circumcircles are, therefore, also congruent. Since the circumcircle of PBC is the circle with center Q and the circumcircle of P^*CB is the circumcircle of ABC, we see that the radius of the circumcircle of ABC is also equal to r, completing the proof of part c).

Finally, to prove part d), we consider Figure 101d-P. We have just shown $AP \perp BC$, and since AP is the common chord of the circles with centers R and S, both R and S lie on the perpendicular bisector of AP, yielding $AP \perp RS$. This gives us $BC \| RS$, and since we can analogously determine $CA \| SQ$ and $AB \| QR$, we see that triangles ABC and QRS are similar. Since we already know that they both have the same circumradius r, $\triangle ABC \cong \triangle QRS$, which completes the proof of part d).

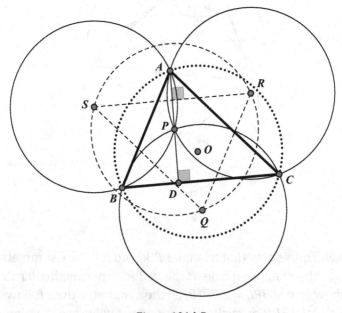

Figure 101d-P

Curiosity 102. Further Unexpected Concurrencies

To approach the proof, we first need to establish a few preliminary results. In Figure 102a-P, the feet of the perpendiculars from point P

on the triangle sides AB, BC, and CA are points G, H, and J, respectively. Applying the Pythagorean theorem to the right triangles $\triangle APJ$, $\triangle BPG$, and $\triangle CPH$, we get $AP^2 + BP^2 + CP^2 = (AJ^2 + JP^2) + (BG^2 + GP^2) + (CH^2 + HP^2)$.

Then applying the Pythagorean theorem to the right triangles $\triangle APG$, $\triangle BPH$, and $\triangle CPJ$, we get

$$AP^2 + BP^2 + CP^2 = (AG^2 + GP^2) + (BH^2 + HP^2) + (CJ^2 + JP^2),$$

which is equivalent to $AJ^2 + CH^2 + BG^2 = AG^2 + BH^2 + CJ^2$. Note that P is in the interior of the triangle in this figure, but the argument also works when P is outside the triangle.

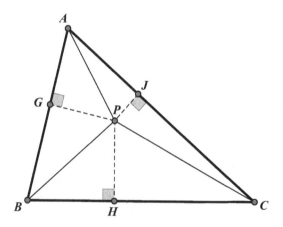

Figure 102a-P

Now that we have established this, we note that the converse is also true. In other words, if points G, H, and J are chosen on the sides AB, BC, and CA of a triangle ABC, respectively, such that $AJ^2 + CH^2 + BG^2 = AG^2 + BH^2 + CJ^2$, and the lines perpendicular to the relevant sides are drawn in each of these points, these three perpendiculars must meet in a common point P. (We restrict ourselves to the case where the points are in the interior of the sides.) This can be shown in the following way. Consider the lines perpendicular to AB in G and perpendicular to CA in J, and assume that these intersect in a point P.

If we draw the line perpendicular to BC through P, it will intersect with BC in a point H', and by the result we have just established, we have $AJ^2 + CH'^2 + BG^2 = AG^2 + BH'^2 + CJ^2$. On the other hand, we have also assumed that $AJ^2 + CH^2 + BG^2 = AG^2 + BH^2 + CJ^2$ holds. This means that $0 = AJ^2 + CH'^2 + BG^2 - (AG^2 + BH'^2 + CJ^2) = AJ^2 + CH^2 + BG^2 - (AG^2 + BH^2 + CJ^2)$, which is equivalent to $CH'^2 - BH'^2 = CH^2 - BH^2$. If we can show that $H = H'$ must follow from this, we will have shown that the three perpendiculars meet in P, as claimed. This follows from the fact that $CH'^2 - BH'^2 = CH^2 - BH^2$ is equivalent to $(CH' - BH')(CH' + BH') = (CH - BH)(CH + BH)$. Since we are assuming that H and H' are between B and C, we have $CH' + BH' = CH + BH = BC$, and from this we obtain $CH' - BH' = CH - BH$. If we now assume $CH' < CH$, this implies $BH' > BH$; therefore, $CH' - BH' < CH - BH$. Similarly, $CH' > CH$ implies $CH' - BH' > CH - BH$, and we see that $CH' = CH$; thus $H' = H$ must hold true. We are now ready to apply these results to the problem at hand.

We apply this relationship in Figure 102b-P to triangle DEF and point Q with its perpendiculars to each of the three sides at points L, M, and N. We need to prove $DL^2 + FN^2 + EM^2 = DM^2 + EN^2 + FL^2$, or $(DL^2 - FL^2) + (FN^2 - EN^2) + (EM^2 - DM^2) = 0$.

Because of the right angles in the figure, the points L, M, and N are certainly interior points of the sides of triangle DEF. In triangle ADF, the Pythagorean theorem gives us $AD^2 - DL^2 = AF^2 - FL^2$ or $DL^2 - FL^2 = AD^2 - AF^2$. Similarly, triangle BED gives us $EM^2 - DM^2 = BE^2 - BD^2$, and triangle CFE gives us $FN^2 - EN^2 = CF^2 - CE^2$. The equation we need to prove is thus equivalent to $(AD^2 - AF^2) + (CF^2 - CE^2) + (BE^2 - BD^2) = 0$. This can also be expressed as $(AD^2 - BD^2) + (BE^2 - CE^2) + (CF^2 - AF^2) = 0$, and this now follows from considering the triangles ABD, BCE, and CAF. Because of the way G, H, and J were defined, we know that $(AG^2 - BG^2) + (BH^2 - CH^2) + (CJ^2 - AJ^2) = 0$ certainly holds true. In triangle ABD we have $GD^2 = AD^2 - AG^2 = BD^2 - BG^2$, which yields $AD^2 - BD^2 = AG^2 - BG^2$. Similarly, from triangle BCE we obtain $BE^2 - CE^2 = BH^2 - CH^2$, and from triangle CAF we obtain $CF^2 - AF^2 = CJ^2 - AJ^2$. This means that $(AD^2 - BD^2) + (BE^2 - CE^2) + (CF^2 - AF^2) = 0$ is equivalent to $(AG^2 - BG^2) + (BH^2 - CH^2) + (CJ^2 - AJ^2) = 0$, and the lines AL, BM, and CN do indeed meet in a common point Q, as we had set out to prove.

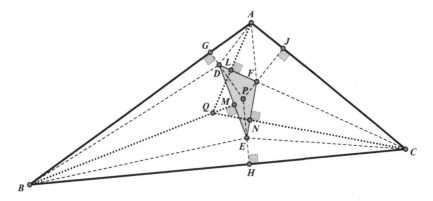

Figure 102b-P

Curiosity 103. One Concurrency Generates Another Concurrency

Since *AD*, *BE*, and *CF* intersect at point *P*, as shown in Figure 103-P, Ceva's theorem (see Toolbox) gives us $\dfrac{BD}{CD} \cdot \dfrac{CE}{AE} \cdot \dfrac{AF}{BF} = 1$. The secants

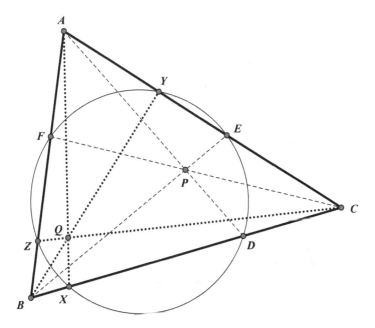

Figure 103-P

AC and *AB* are drawn from an external point when they intersect the circle we have *AY·AE = AF·AZ*, which is equivalent to $\dfrac{AF}{AE} = \dfrac{AY}{AZ}$, and analogously, for vertices *B* and *C* we also obtain $\dfrac{BD}{BF} = \dfrac{BZ}{BX}$ and $\dfrac{CE}{CD} = \dfrac{CX}{CY}$.

Substituting these expressions in $\dfrac{BD}{CD} \cdot \dfrac{CE}{AE} \cdot \dfrac{AF}{BF} = 1$, which we can also write as $\dfrac{BD}{BF} \cdot \dfrac{CE}{CD} \cdot \dfrac{AF}{AE} = 1$, we thus obtain $\dfrac{BZ}{BX} \cdot \dfrac{CX}{CY} \cdot \dfrac{AY}{AZ} = 1$, which we can write as $\dfrac{CX}{BX} \cdot \dfrac{AY}{CY} \cdot \dfrac{BZ}{AZ} = 1$. By the converse of Ceva's theorem, we find that *AX*, *BY*, and *CZ* are also concurrent, as we sought to prove.

Curiosity 104. Unusual Perpendiculars Generating Concurrencies

In Figure 104-P, we have added the midpoints *X*, *Y*, and *Z* of line segments *DK*, *EL*, and *FM*, respectively, as well as the perpendicular

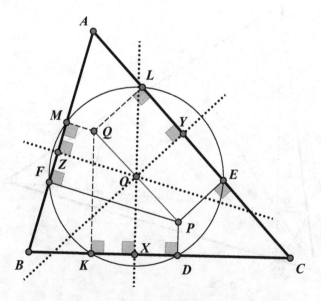

Figure 104-P

bisectors OX, OY, and OZ of line segments DK, EL, and FM, respectively, which intersect at the center O of the original circle. When PQ is drawn, forming trapezoid $QKDP$, point O is the midpoint of PQ. Similarly, for trapezoid $QMFP$, we once again have point O as the midpoint of QP. Therefore, we have $PO = QO$. Furthermore, we have the midpoint O of side QP of trapezoid $ELQP$. Thus, we see that the perpendiculars QK, QL, and QM will be concurrent at point Q, since point Q enables the other concurrency.

Curiosity 105. A Most Unexpected Concurrency

In Figure 105-P, we shall first consider triangle KMN. Since M, N, and G are the midpoints of AC, AB, and AD, respectively, we have $MN \| BC$, so that $\triangle ADC \sim \triangle AGM$ and $\triangle ANG \sim \triangle ABD$ and then $\dfrac{NG}{BD} = \dfrac{AG}{GD} = \dfrac{MG}{CD}$,

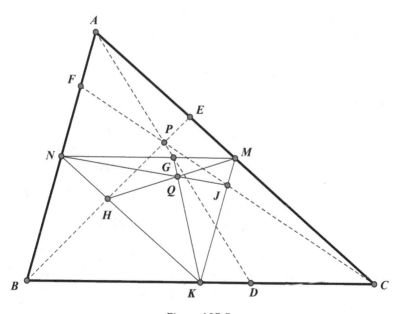

Figure 105-P

which then can be written as $\dfrac{BD}{CD} = \dfrac{NG}{MG}$. Analogously, we also obtain $\dfrac{CE}{AE} = \dfrac{KH}{NH}$ and $\dfrac{AF}{BF} = \dfrac{MJ}{KJ}$.

Since AD, BE, and CF are concurrent, Ceva's theorem (see Toolbox) gives us $\dfrac{BD}{CD} \cdot \dfrac{CE}{AE} \cdot \dfrac{AF}{BF} = 1$. Substituting for each of the three fractions, we get $\dfrac{NG}{MG} \cdot \dfrac{KH}{NH} \cdot \dfrac{MJ}{KJ} = 1$, and KG, MH, and NJ are thus concurrent by Ceva's theorem.

Curiosity 106. An Intriguing Concurrency

In Figure 106-P, lines AK, BM, and CN have been extended to intersect lines BC, CA, and AB at X, Y, and Z, respectively. Since we want to show that these three extensions are concurrent, we would like to apply Ceva's theorem (see Toolbox) and show that $\dfrac{BX}{CX} \cdot \dfrac{CY}{AY} \cdot \dfrac{AZ}{BZ} = 1$.

As a first step, we note that in triangle ABX, by the law of sines (see Toolbox), we have $\dfrac{BX}{AX} = \dfrac{\sin\angle BAX}{\sin\angle B}$. Note that we will, as usual, use $\angle A$, $\angle B$, and $\angle C$ to denote the angles at the vertices of triangle ABC. Similarly, in $\triangle AXC$ we have $\dfrac{CX}{AX} = \dfrac{\sin\angle XAC}{\sin\angle C}$, and dividing the first of these equalities by the second gives us $\dfrac{BX}{CX} = \dfrac{\sin\angle C \cdot \sin\angle BAX}{\sin\angle B \cdot \sin\angle XAC}$.

Now we apply the law of sines to triangle AFK. Here, we obtain $\dfrac{AF}{FK} = \dfrac{\sin\angle AKF}{\sin\angle BAX}$, which can be rewritten as $AF \sin \angle BAX = FK \sin \angle AKF$. Similarly, in triangle AKE, we obtain $\dfrac{AE}{EK} = \dfrac{\sin\angle EKA}{\sin\angle XAC}$, which can be rewritten as $AE \sin \angle XAC = EK \sin \angle EKA$. Since $FK = EK$ and $\sin \angle AKF = \sin \angle EKA$, we get $AF \sin \angle BAX = AE \sin \angle XAC$, which can be rewritten as $\dfrac{AE}{AF} = \dfrac{\sin\angle BAX}{\sin\angle XAC}$. Substituting, we get

$$\frac{BX}{CX} = \frac{\sin\angle C \cdot \sin\angle BAX}{\sin\angle B \cdot \sin\angle XAC} = \frac{\sin\angle C}{\sin\angle B} \cdot \frac{AE}{AF}.$$ Analogously, we also obtain

$$\frac{CY}{AY} = \frac{\sin\angle A}{\sin\angle C} \cdot \frac{BF}{BD} \text{ and } \frac{AZ}{BZ} = \frac{\sin\angle B}{\sin\angle A} \cdot \frac{CD}{CE}, \text{ and this gives us}$$

$$\frac{BX}{CX} \cdot \frac{CY}{AY} \cdot \frac{AZ}{BZ} = \frac{\sin\angle C}{\sin\angle B} \cdot \frac{AE}{AF} \cdot \frac{\sin\angle A}{\sin\angle C} \cdot \frac{BF}{BD} \cdot \frac{\sin\angle B}{\sin\angle A} \cdot \frac{CD}{CE} = \frac{AE}{AF} \cdot \frac{BF}{BD} \cdot \frac{CD}{CE} = 1,$$

by applying Ceva's theorem. This proves that AK, BM, and CN are indeed concurrent.

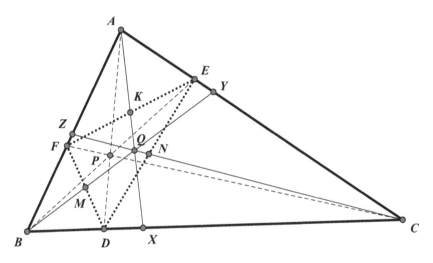

Figure 106-P

Curiosity 107. Another Unexpected and Unforeseen Concurrency

We know that the median to the hypotenuse of a right triangle is half the length of the hypotenuse. Consequently, when two right triangles share the same hypotenuse, their medians to the hypotenuse are equal. Therefore, for right triangles BCQ and BCR in Figure 107-P, which share the hypotenuse BC, we have $QD = RD$. We then have DX as the perpendicular bisector of RQ. We can make the same argument to show that EY is the perpendicular bisector of RP and that FZ the perpendicular bisector of QP. The perpendicular bisectors of the three sides of a triangle (in this case, $\triangle PQR$) meet at the center of the circumscribed circle of that triangle and are, therefore, concurrent.

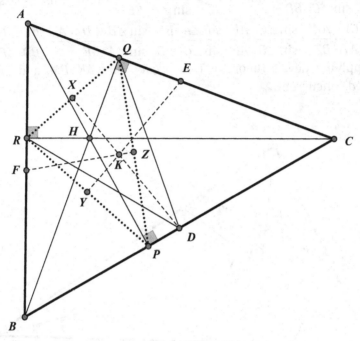

Figure 107-P

Curiosity 108. A Most Unexpected Concurrency from a Triangle

In Figure 108-P, we first extend PD to intersect AB at point N. Applying Ceva's theorem (see Toolbox), we get $\dfrac{AN}{BN} \cdot \dfrac{BC}{PC} \cdot \dfrac{PE}{AE} = 1$.

We also have $EX \| AB$, so that $\dfrac{BC}{PC} = \dfrac{AE}{PE}$. By substitution, this gives us

$\dfrac{AN}{BN} \cdot \dfrac{AE}{PE} \cdot \dfrac{PE}{AE} = 1$, which reduces to $\dfrac{AN}{BN} = 1$. This is equivalent to $AN = BN$, which means that point N is the midpoint of AB, and thus equal to point M. Therefore, the line PD (extended) goes through point M, and this justifies the concurrency at point P.

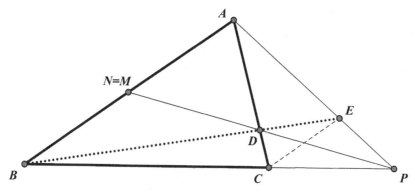

Figure 108-P

Curiosity 109. Astounding Point Property in a Triangle

We once again apply Ceva's theorem (see Toolbox) to show that AD, BE, and CF meet in a common point. We must show $\dfrac{BD}{CD} \cdot \dfrac{CE}{AE} \cdot \dfrac{AF}{BF} = 1$.

Since PD was constructed as the bisector of $\angle BPC$, we have $\dfrac{BD}{CD} = \dfrac{PB}{PC}$ (see Toolbox), and similarly for angles $\angle CPA$ and $\angle APB$ we obtain $\dfrac{CE}{AE} = \dfrac{PC}{PA}$ and $\dfrac{AF}{BF} = \dfrac{PA}{PB}$, respectively. Appropriately substituting gives us $\dfrac{BD}{CD} \cdot \dfrac{CE}{AE} \cdot \dfrac{AF}{BF} = \dfrac{PB}{PC} \cdot \dfrac{PC}{PA} \cdot \dfrac{PA}{PB} = 1.$

By Ceva's theorem, we see that AD, BE, and CF meet in a common point. We now need to show that the inequalities $\frac{1}{2}(AB + BC + CA) < PA + PB + PC < 2(AB + BC + CA)$ hold true.

In triangle PBC, the triangle inequality gives us $PB + PC > BC$. Similarly, in triangles PCA and PAB, we obtain $PC + PA > CA$ and $PA + PB > AB$, respectively. Adding these three inequalities gives us $2PA + 2PB + 2PC > AB + BC + CA$, which is equivalent to the left-side inequality $\frac{1}{2}(AB + BC + CA) < PA + PB + PC$.

To prove the right-side inequality $PA + PB + PC < 2(AB + BC + CA)$, we define point X as the intersection of BC with AP extended, as shown in Figure 109-P. Since P is in the interior of ABC, we have $PA < XA$ and

$BX < BC$. The triangle inequality in $\triangle ABX$ gives us $PA < XA < AB + BX <$ $AB + BC$. Since we can proceed analogously for PB and PC, we also have $PB < BC + CA$ and $PC < CA + AB$, and adding these three inequalities gives us the right-side inequality $PA + PB + PC < 2(AB + BC + CA)$.

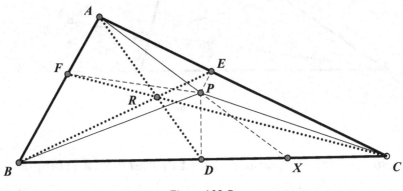

Figure 109-P

Curiosity 110. A Counterintuitive Concurrency

In Figure 110-P, we define the intersection of EH with DG as point M and the intersection of FJ with DG as point N. It is our goal to show that $M = N$, as this then indicates that lines DG, EH, and FJ are concurrent in the common point $M = N = P$. We also have the intersection of AD with BC as point K, the intersection of AD with EH as point L, and the intersection of EH with BC as point Q. With the perpendiculars indicated in Figure 110-P, we find that $\angle MED = \angle CAK$, because they are each complementary to the vertical angles at point L in triangles DLE and HAL, respectively. We also have angle ACQ complementary to angle CQH, and angle QMG complementary to angle CQH; therefore, $\angle ACK = \angle ACQ = \angle QMG = \angle EMD$. Triangles MED and CAK are therefore, similar and we have $\dfrac{DM}{DE} = \dfrac{KC}{KA}$. In a similar fashion, we can show that $\triangle BKA \sim \triangle FND$, so that $\dfrac{DF}{DN} = \dfrac{KA}{KB}$. Because of the perpendicularity, we have $BE\|AD\|CF$, which enables us to set up the proportion $\dfrac{DE}{DF} = \dfrac{KB}{KC}$. When we multiply these last three proportions, we get

$$\frac{DM}{DE} \cdot \frac{DF}{DN} \cdot \frac{DE}{DF} = \frac{KC}{KA} \cdot \frac{KA}{KB} \cdot \frac{KB}{KC}, \text{ which reduces to } \frac{DM}{DN} = 1, \text{ or } DM = DN.$$

This implies that points M and N coincide, thus proving the concurrence of the three lines DG, EH, and FJ.

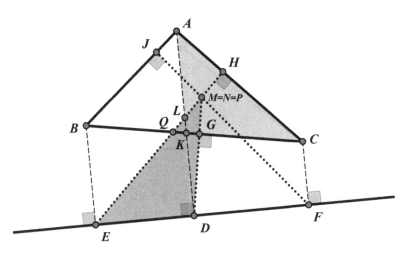

Figure 110-P

Curiosity 111. Concurrent Angle Bisectors of Triangles with a Common Base

Since the angles at points C, D, E, and F are of equal measure, they are concyclic on the common chord AB of the circumcircle of triangle ABC, as shown in Figure 111-P. Each of these angles has a measure one-half of the measure of arc AB. The bisectors of each of these angles will also meet at the midpoint P of arc AB; therefore, the bisectors are concurrent.

Curiosity 112. An Unexpected Concurrency with the Circumscribed Circle

In Figure 112-P, we have point P as the intersection of KL and AO. Our objective is to show that $PD \perp BC$. Triangle PAD is isosceles, since KL

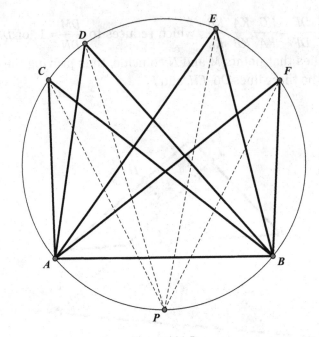

Figure 111-P

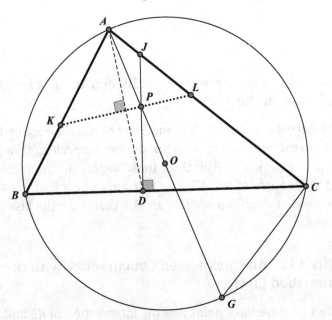

Figure 112-P

is the bisector of AD; therefore, $\angle DAP = \angle PDA = x$. Let $\angle CAP = y$. Since AD bisects $\angle BAC$, we have $\angle DAB = \angle CAD = x + y$. Then let $\angle ADB = z$, and define point G as diametrically opposite to point A on the circumscribed circle of triangle ABC, as shown in Figure 112-P. We note that $\angle ACG = 90°$ because AG is a diameter of the circle; therefore, $\angle CBA = \angle AGC = 90° - y$. In $\triangle ABD$, we then have $180° = \angle CBA + \angle BAD + \angle ADB = (90° - y) + (x + y) + z = 90° + x + z$, from which we obtain $\angle JDB = \angle JDA + \angle ADB = x + z = 90°$; thus, $PD \perp BC$. This justifies the concurrency of lines KL, DJ, and AO, which intersect at point P.

Curiosity 113. Another Surprising Aspect of Concurrent Cevians

This configuration has lots of parallelograms based on the construction. We begin Figure 113-P by noticing that quadrilaterals $FNDU$ and $FUEM$ are both parallelograms, since both pairs of opposite sides are

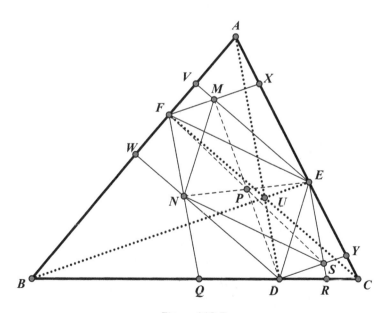

Figure 113-P

parallel. Consequently, we know that *ND* = *ME*. Therefore, *MNDE* is also a parallelogram, whose diagonals *NE* and *MD* bisect each other at point *P*. Furthermore, quadrilateral *DSEU* is also a parallelogram, and thus *FN* = *UD* = *ES*. This shows us that quadrilateral *FNSE* is also a parallelogram, with its diagonals *NE* and *FS* also bisecting each other at the common point *P*. Therefore, we have shown that the three lines *MD*, *NE*, and *SF* are concurrent at point *P*, as they bisect each other.

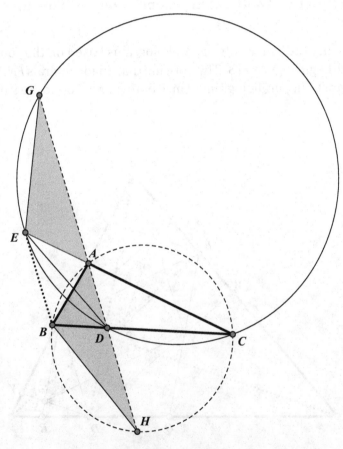

Figure 114-P

Curiosity 114. A Surprising Equality

Consider the triangles AGE and ABH in Figure 114-P. First, however, a few preliminary observations are useful. We will define $\angle BAC = \alpha$, and since AD bisects this angle, we have $\angle BAD = \angle DAC = \dfrac{\alpha}{2}$. This angle will be considered a few more times. Since they are vertical angles, we have $\angle GAE = \angle DAC = \dfrac{\alpha}{2}$, and since BE is parallel to AD, we also have $\angle BAD = \angle ABE = \dfrac{\alpha}{2}$. Finally, in triangle AEB, we also have $\angle BEA = \angle BAC - \angle ABE = \alpha - \dfrac{\alpha}{2} = \dfrac{\alpha}{2}$. This determines that triangle AEB is isosceles with $\angle ABE = \angle BEA = \dfrac{\alpha}{2}$. We also have $\angle EGA = \angle EGD = \angle ECD = \angle ACB$, and in the circumscribed circle of triangle ABC, we have $\angle ACB = \angle AHB$. Therefore, $\angle EGA = \angle AHB$. Now we are ready to return to the triangles AGE and ABH, where $\angle GAE = \angle BAH = \dfrac{\alpha}{2}$ and $\angle EGA = \angle AHB$. This indicates that the triangles AGE and ABH are congruent, and then $AG = AH$, which is what we sought to prove.

Curiosity 115. An Unforeseen Triangle Surprise

We begin by drawing some auxiliary lines in Figure 115-P. We extend BK to point D, so that $BK = DK$, and we extend CH to point E, so that $CH = EH$. We then draw line segments AD, CD, AE, and BE. Since AH is a perpendicular bisector of CE, we find that $\triangle AHC \cong \triangle AEH$; therefore, $AC = AE$, and $\angle CAH = \angle HAE$. In a similar fashion, we can show that $\triangle AKD \cong \triangle ABK$ with $AD = AB$, and also $\angle DAK = \angle KAB$. By subtraction we then obtain: $\angle BAE = \angle BAK - \angle EAH = \angle KAD - \angle HAC = \angle CAD$. We then can conclude that $\triangle DAC \cong \triangle BEA$, and thus $CD = BE$. We now consider $\triangle DBC$, where points K and M are the midpoints of sides BD and BC, respectively. Therefore, $MK = \frac{1}{2} \cdot CD$. Similarly, in $\triangle CBE$, points M and H are midpoints of sides CB and CE, respectively, and we have $MH = \frac{1}{2} BE$. Since $CD = BE$, we thus have $MK = MH$.

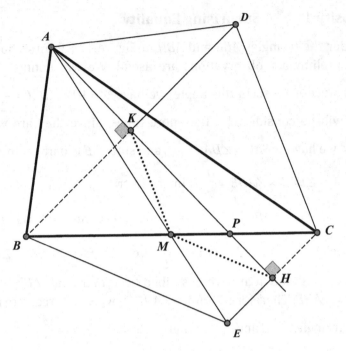

Figure 115-P

Curiosity 116. A Remarkable Property of Two Triangles with a Common Base

In Figure 116a-P, we have added the common altitude AR of triangles ABC and DBC. Line segment AR intersects EH in point Q, and AD is the altitude of triangles AEF and DGH. Since $BC\|EF$, triangles ABC and AEF are similar, and we have $\dfrac{EF}{BC}=\dfrac{AQ}{AR}$. Now, considering triangles DBC and DGH, these are also similar, with $BC\|GH$, and we have $\dfrac{GH}{BC}=\dfrac{AQ}{AR}$. This gives us $\dfrac{EF}{BC}=\dfrac{AQ}{AR}=\dfrac{GH}{BC}$; thus, $EF=GH$. This argument is independent of the distance between EH and BC and therefore also holds for IL. We then also obtain $IK=JL$, and it follows that $IJ+JK=JK+KL$, which is equivalent to $IJ=KL$.

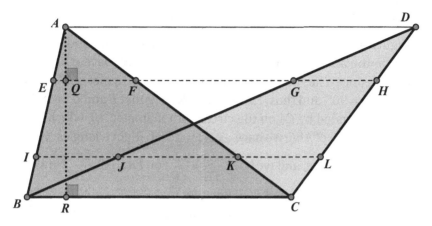

Figure 116a-P

Considering the situation illustrated in Figure 116b-P, we see that an analogous argument to the one given above yields $\dfrac{MN}{BC} = \dfrac{AQ}{AR} = \dfrac{OP}{BC}$; thus $MN = OP$. From this we also obtain $MP + PN = OM + MP$, which yields $PN = OM$.

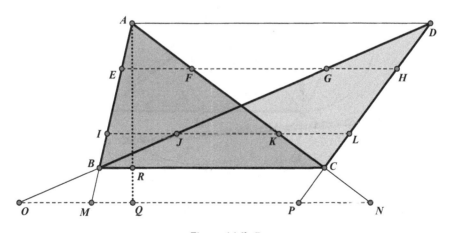

Figure 116b-P

Curiosity 117. A Surprising Line Partitioning

In Figure 117-P, we have added line segments *BE*, *AF*, and *AG*. Since *E* lies on the arc subtended on the diameter *BC*, we have $\angle BEC = 90°$. Similarly, since *F* lies on the arc subtended on the diameter *CA*, we also have $\angle CFA = 90°$, and thus, $\angle AFG = 90°$. Noting that *B* and *G* both lie on the arc subtended by *CA* on the circle with diameter *AB*, which is also the circumcircle of *ABC*, we have $\angle CGA = \angle CBA$. Right triangles *FAG* and *CAB* are thus similar, and we have $\dfrac{FA}{FG} = \dfrac{CA}{CB}$, or $FA = FG \cdot \dfrac{CA}{CB}$. Now, considering triangles *EBC* and *FCA*, we note that $\angle ACF = \angle ACB - \angle ECB = 90° - \angle ECB = \angle CBE$. This means that right triangles *EBC* and *FCA* are also similar, which implies $\dfrac{CE}{CB} = \dfrac{FA}{CA}$, or $CE = FA \cdot \dfrac{CB}{CA}$. Appropriate substitution then gives us $CE = FG \cdot \dfrac{CA}{CB} \cdot \dfrac{CB}{CA} = FG$.

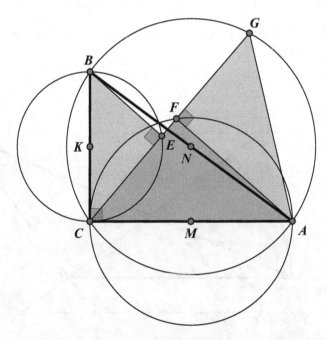

Figure 117-P

Curiosity 118. The Hidden Length Equality of Line Segments in a Triangle

In Figure 118-P, we notice immediately that quadrilateral $ADHE$ is a parallelogram, since the opposite sides are parallel. The diagonals AH and DE bisect each other at point M. Since $DF\|AH\|EG$, we find that AH also bisects FG, so that $FH = GH$. We now define point K as the intersection of AH with the line parallel to BC through E. Since $EKHG$ is then also a parallelogram, we have $FH = GH = EK$. Since we also have corresponding angles of parallel lines $\angle DHF = \angle AEK$ and $\angle HFD = \angle EKA$, we have $\triangle AKE \cong \triangle DFH$. From this, we obtain $AK = DF$. Reverting again to parallelogram $EKHG$, where $KH = EG$, we have $DF + EG = AK + KH = AH$.

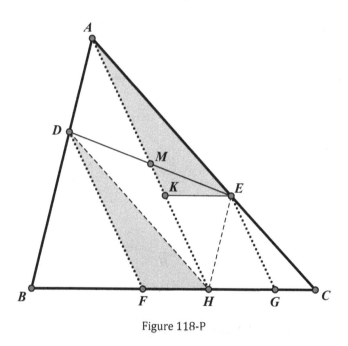

Figure 118-P

Curiosity 119. The Unexpected Angle Measure

When we circumscribe a circle around triangle ABC, as shown in Figure 119-P, and extend the lines CD, CF, and CE, they meet the circumscribed circle at points G, H, and J, respectively. Since the angles at point *C* are equal, the arcs *AG*, *GH*, *HJ*, and *JB* are also equal. Therefore, we have *GJ*∥*AB*, and since *AB* is perpendicular to *CG*, we know that ∠*CGJ* is a right angle. The segment *CEJ* is therefore a diameter of the circle. Since the point *E* is the midpoint of *AB*, it must be the circumcenter of triangle *ABC* and *AB* is then a diameter of the circle. We see that ∠*ACB* is indeed a right angle.

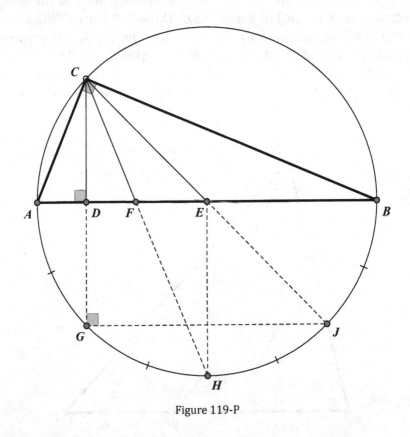

Figure 119-P

Curiosity 120. An Unexpected Equality from a Right Triangle

Proving triangles *CAQ* and *PCB* congruent will bring us the desired result that *PB* = *QC*, as shown in Figure 120-P. First, we note ∠*CAQ* = 180° − ∠*DAC*. Since △*CAD* ∼ △*BCD*, we have ∠*DAC* = ∠*BCD*; therefore, ∠*CAQ* = 180° − ∠*BCD*. Since ∠*BCP* = 180° − ∠*BCD*, we therefore have

$\angle CAQ = \angle BCP$. Because we are given that $CP = AC$ and $AQ = BC$, this implies $\triangle CQA \cong \triangle BPC$, and thus $PB = QC$.

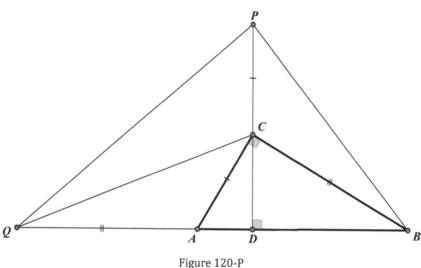

Figure 120-P

Curiosity 121. The Conundrum: Perpendicular or Parallel

In Figure 121a-P, with BP oriented to the right, we calculate $\angle BPC = 90° - \angle PNB = 90° - \angle CNA$. Since $BP = BC$, we therefore have $\angle PCB = \angle BPC = 90° - \angle CNA$, and thus $\angle PCB = 90° - \angle CNA = 90° - \angle ACN$. Therefore, $\angle CNA = \angle ACN$, and it follows that $AC = AN$. In triangles NEA and CAE, with the common side AE, we therefore have $AN = AC$ and $\angle NAE = \angle EAC$, which enables us to show $\triangle NAE \cong \triangle EAC$, so that $\angle CAE = \angle AEN$, which are two equal angles whose sum is 180°. Each is therefore equal to 90°, so that we have proved that $AC \perp PC$.

In Figure 121b-P, on the other hand, with BP oriented to the left, we calculate $\angle AFB = 90° - \angle BAF = 90° - \frac{1}{2}\angle BAC$ in right triangle ABF. For right triangle ABC, we have $\angle CDA = 90° - \angle DAC = 90° - \frac{1}{2}\angle BAC$; therefore, $\angle AFB = \angle CDA = \angle BDF$, from which we obtain $BF = BD$. Since $BP = BC$, we have $\dfrac{BF}{BP} = \dfrac{BD}{BC}$. Thus, $DF \| CF$ and $FA \| CP$.

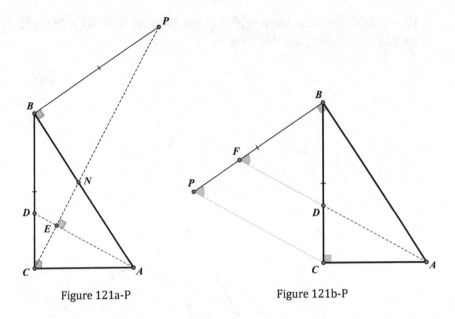

Figure 121a-P   Figure 121b-P

Curiosity 122. Multiple Midpoints in a Right Triangle Determine Equal Line Segments

In Figure 122-P, we begin the proof by drawing lines *FG, GJ, JH,* and *HF.* In triangle *DEB*, the line *FG* joins the midpoints of two sides *BE*

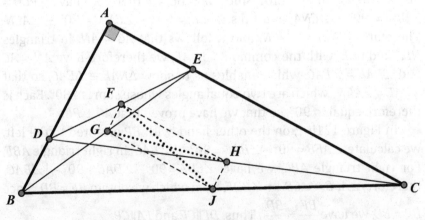

Figure 122-P

and *DE*, and is therefore equal to half the length of the third side and parallel to it, so that $FG = \frac{1}{2}DB$ and $FG\|DB$. Similarly, in triangle *DBC*, *HJ* joins the midpoints of two sides of the triangle, so that we have $HJ = \frac{1}{2}DB$ and $HJ\|DB$. Therefore, $FG = HJ$ and $FG\|HJ$. Thus, we can conclude that quadrilateral *FGJH* is a parallelogram because $AB \perp AC$ with $FG\|AB$ and $FH\|AC$, since *FH* joins the midpoints of two sides of triangle *EDC*. Therefore, $FG \perp FH$. That establishes that parallelogram *FGJH* is a rectangle, which has equal diagonals, so we can conclude that $GH = FJ$.

Curiosity 123. Perpendiculars in Right Triangles that Generate Equal Angles

We first note that $\angle CBA = 90° - \angle BAC = \angle ACD - \angle ECD$, which can be seen in Figure 123-P. This means that right triangles *ABC* and *CED* are similar with $\dfrac{DE}{CE} = \dfrac{CA}{CB}$. Since $DE = CF$, this is equivalent to $\dfrac{CF}{CE} = \dfrac{CA}{CB}$ or $\dfrac{CF}{CA} = \dfrac{CE}{CB}$, and right triangles *CAF* and *CEB* are therefore also similar. This implies $\angle CFA = \angle BEC$, and thus $\angle AEB = 180° - \angle BEC = 180° - \angle CFA = \angle AFB$, as we set out to prove.

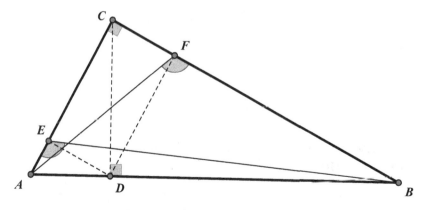

Figure 123-P

Curiosity 124. Right Triangles Sharing a Common Hypotenuse Generate an Unexpected Equality

In Figure 124-P, we have added line segment *MP*, and we will show that triangles *PCM* and *PMD* are congruent. Notice that *CM* and *DM* are medians to the common hypotenuse in right triangles *ACB* and *ABD*, respectively. We know that the median to the hypotenuse of a right triangle is half the length of the hypotenuse; thus, $CM = \frac{1}{2}AB = DM$. This means that points *A*, *C*, *B*, and *D* all lie on a common circle with center *M*, as shown in Figure 124-P. Since right triangles *PCM* and *PMD* share the hypotenuse *PM* and we have *CM = DM*, we see that they are congruent, which implies *PC = PD*.

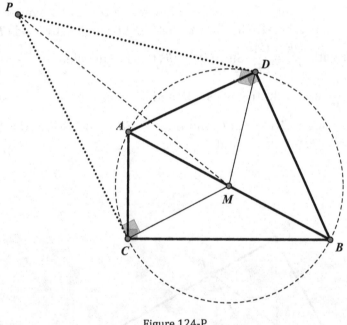

Figure 124-P

Curiosity 125. The Pythagorean Theorem Revisited Geometrically

There are currently well over 400 proofs of the Pythagorean theorem and the number continues to grow. It is particularly noteworthy that one of these proofs was developed by the 20th president of the United States, James A. Garfield. While a member of the House of Representatives, Garfield, who enjoyed "playing" with elementary mathematics, stumbled upon a cute proof of this famous theorem.[6] It was subsequently published in the *New England Journal of Education* after encouragement by two professors (Quimby and Parker) at Dartmouth College, where he had given a lecture on March 7, 1876.

Garfield's proof is actually quite simple, as it begins by placing two congruent right triangles *ABE* and *ECD* so that points *B*, *C*, and *E* are collinear as shown in Figure 125-P. Because of the right angles in *B* and *C*, we note that this forms a trapezoid *CDAB*. Notice also that since $\angle AEB + \angle CED = 90°$, we obtain $\angle DEA = 180° - (\angle AEB + \angle CED) = 180° - 90° = 90°$, making *AED* a right isosceles triangle with *AE = DE*.

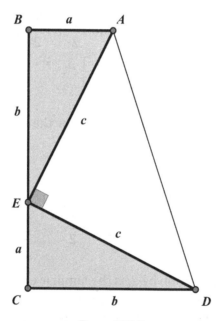

Figure 125-P

---

[6]In October 1851 he noted in his diary that "I have today commenced the study of geometry alone without class or teacher".

We can now calculate the area of trapezoid $ABCD$ in two different ways. First, using the usual formula to calculate the area of a trapezoid, we have

$$\text{area}[ABCD] = \frac{1}{2}(\text{sum of bases})(\text{altitude})$$

$$= \frac{1}{2}(a+b)(a+b)$$

$$= \frac{1}{2}a^2 + ab + \frac{1}{2}b^2.$$

On the other hand, by calculating the sum of the areas of the three triangles that compose the trapezoid, we obtain

$$\text{area}[ABCD] = \frac{1}{2}ab + \frac{1}{2}ab + \frac{1}{2}c^2$$

$$= ab + \frac{1}{2}c^2.$$

We now equate the two expressions for the area of the trapezoid to get:

$$\frac{1}{2}a^2 + ab + \frac{1}{2}b^2 = ab + \frac{1}{2}c^2$$

$$or \quad \frac{1}{2}a^2 + \frac{1}{2}b^2 = \frac{1}{2}c^2,$$

which, after multiplication by 2, is the familiar expression $a^2 + b^2 = c^2$, also known as the Pythagorean theorem.

As mentioned earlier, there are over 400 proofs[7] of the Pythagorean theorem available today; many are ingenious, yet some are a bit

---

[7]A classic source for 370 proofs of the Pythagorean theorem is Elisha S. Loomis' *The Pythagorean Proposition* (Reston, VA: NCTM, 1968).

cumbersome. However, it is interesting to note that none will ever use trigonometry to prove the Pythagorean theorem. There can be no proof of the Pythagorean theorem using trigonometry, since the whole discipline of trigonometry is based on the validity of the Pythagorean theorem. Thus, any such proof using trigonometry would mean proving the very theorem on which trigonometry depends, resulting in a typical case of circular reasoning.

Curiosity 126. Squares on Triangle Sides Produce Noteworthy Surprises

There are various aspects to be proved here so we will partition the discussion into several sections. Note that there are three claims to be proved. First, we claim that lines *AD*, *BE*, and *CF* are concurrent. Then, we claim that each of these lines is perpendicular to a side of *DEF*, where $AD \perp EF$, $BE \perp DF$, and $CF \perp DE$. Finally, we state that the perpendicular segments are also of equal length so that $AD = EF$, $BE = DF$, and $CF = DE$, as illustrated in Figure 126a-P.

If we can show that the lines *AD*, *BE*, and *CF* are perpendicular to *BC*, *CA*, and *AB*, respectively, this would mean that they are the altitudes of $\triangle ABC$; therefore, they would intersect at the orthocenter of $\triangle ABC$. It is then sufficient to prove the other two properties of this configuration.

Part 1. In order to prepare the proof, we draw segments *BR* and *CS* as shown in Figure 126b-P and show that they are perpendicular and of equal length. To do this, we consider the triangles *ABR* and *ASC*. We note that $AR = AC$ and $AB = AS$ and $\angle BAR = \angle BAC + \angle CAR = \angle BAC + 90° = \angle BAC + \angle SAB = \angle SAC$.

This means that triangles *ABR* and *ASC* are congruent. We can now apply a clever trick. Noting that $AR \perp AC$ and $AB \perp AS$, we see that a clockwise rotation of $\triangle ABR$ by 90° will place $\triangle ABR$ congruent to $\triangle ASC$. This means that rotating *BR* about *A* by 90° will result in segment *SC*. This means that the segments *BR* and *CS* are perpendicular and of equal length, as we set out to prove.

Part 2. Next, we let *Z* denote the midpoint of side *BC* as shown in Figure 126c-P. We draw segments *ZE* and *ZF* and show that these are also perpendicular and of equal length.

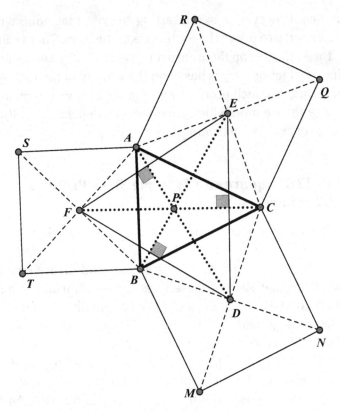

Figure 126a-P

Consider triangles *BZF* and *BCS*. These have a common angle at *B*, and since $\dfrac{BZ}{BC} = \dfrac{BF}{BS} = \dfrac{1}{2}$, they are similar. This means that side *ZF* of $\triangle BZF$ is parallel to side *CS* of $\triangle BCS$, and $\dfrac{BZ}{BC} = \dfrac{ZF}{CS} = \dfrac{1}{2}$. In other words, *ZF* is parallel to *CS* and half as long. We can repeat this argument with triangles *CEZ* and *CRB*, and see that *EZ* is parallel to *RB* and half as long. We have already established that *BR* and *CS* are perpendicular and of equal length, and from this it follows that *ZF* and *ZE* are also perpendicular and of equal length.

Part 3. As we see in figure 126d-P, we are now ready to consider the triangles *ZEB* and *ZFD*. We can now prove that these two triangles are also congruent as follows. We have just established that

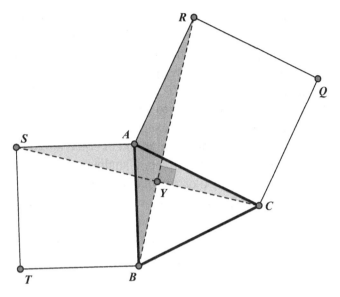

Figure 126b-P

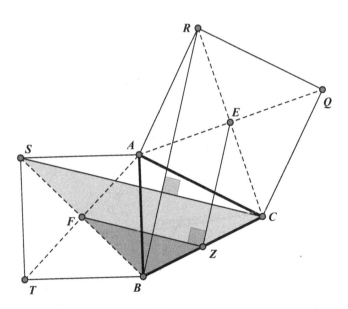

Figure 126c-P

$ZE = ZF$, and both ZB and ZD are half as long as BC, and therefore also of equal length. Furthermore, we have $\angle EZB = \angle EZF + \angle FZB = 90° + \angle FZB = \angle BZD + \angle FZB = \angle FZD$, and we see that triangles ZEB and ZFD are indeed congruent. We can now apply the same rotational trick we used in Part 1. Since $ZE \perp ZF$ and $ZD \perp ZB$, rotation of $\triangle ZEB$ by 90° will let $\triangle ZEB$ be overlapping congruent to $\triangle ZFD$, and this means that rotating BE about Z by 90° will result in segment DF. This means that the segments BE and DF are perpendicular and of equal length.

This can be repeated in a completely analogous way to show that CF and DE are also perpendicular and of equal length, as are AD and EF. This completes the proof, as we have now shown $AD \perp EF$, $BE \perp DF$, and $CF \perp DE$, as well as $AD = EF$, $BE = DF$, and $CF = DE$, and this also implies that AD, BE, and CF intersect in the orthocenter of triangle ABC.

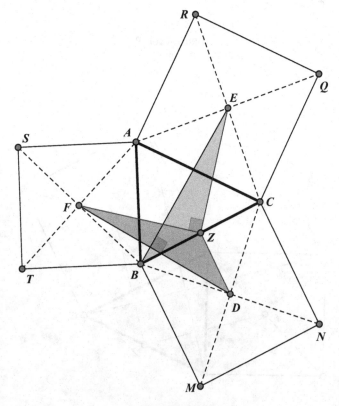

Figure 126d-P

Curiosity 127. More Properties Generated by Squares on Triangle Sides

In Figure 127-P, we define A' on BC as the foot of the perpendicular from vertex A of triangle ABC. With this point, we can then define h as the length of the altitude AA' of $\triangle ABC$ along with $m = BA'$ and $n = CA'$. Furthermore, we define X' as the foot of the perpendicular from X on the extension of AA' and Y' as the foot of the perpendicular from Y on the extension of AA'. In order to prove that triangle AXY is a right isosceles triangle, we will first show that the right triangles AXX' and $AY'Y$ are congruent. Since $AC = QC$, and $\angle ACA' = 180° - \angle DCQ - \angle QCA = 180° - \angle DCQ - 90° = \angle XQD$, we have $\triangle AA'C \cong \triangle CDQ$; therefore, $DQ = A'C = n$, and $CD = AA' = h$. We also have $QY = CN = BC = n + m = a$, and

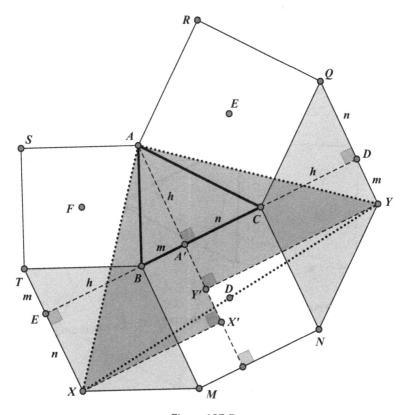

Figure 127-P

$A'Y' = DY = QY - QD = a - n = m$. It follows that $YY' = DA' = n + h$, and $AY' = m + h$. Furthermore, on the other side we also obtain $\triangle ABA' \cong \triangle BET$, $BE = AA' = h$, and $TE = BA' = m$. Then $XX' = A'E = A'B + BE = m + h$. Also, $TX = BM = BC = a$, and $A'X = EX = TX - TE = a - m = n$. Then, $XX' = m + h = AY'$. In summary, this gives us $\triangle AXX' \cong \triangle AYY'$. We then have $AX = AY$, as well as $\angle XAY = \angle XAX' + \angle Y'AY = \angle XAX' + \angle X'XA = 90°$, or $AX \perp AY$, which is what we set out to prove.

Curiosity 128. An Unexpected Perpendicularity

We have already proved this relationship in Part 1 of Curiosity 126, by simply renaming the lines AM and CT as the lines BR and CS.

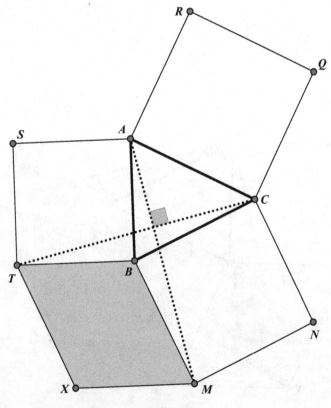

Figure 128-P

Curiosity 129. A Surprising Concurrency from Squares on the Sides of a Triangle

The proof of this result can be derived by a somewhat surprising reconfiguration of the parts of the proof to Curiosity 126. Recalling Part 1 of that proof, we know that BR and CS are perpendicular and of equal length. We can now apply the concept of Part 2 of the proof in a slightly modified way, which is shown in Figure 129a-P. In this instance, we draw segments GE and GF and show that these are also perpendicular and of equal length. We note that G is the midpoint of parallelogram $ARZS$ and, thus the midpoint of segment RS. Now let us consider triangles RGE and RSC. These have a common angle at point R, and since $\dfrac{RG}{RS} = \dfrac{RE}{RC} = \dfrac{1}{2}$, they are similar. This means that side EG of $\triangle RGE$ is parallel to side CS of $\triangle RSC$, and $\dfrac{RG}{RS} = \dfrac{EG}{CS} = \dfrac{1}{2}$. In other words,

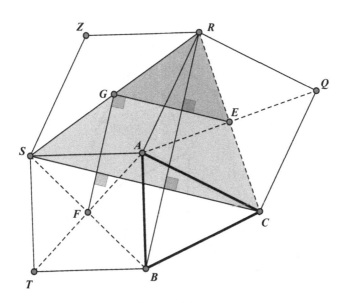

Figure 129a-P

EG is parallel to *CS* and half as long. We can repeat this argument with triangles *SFG* and *SBR*, and see that *GF* is parallel to *RB* and half as long. We have already established that *BR* and *CS* are perpendicular and of equal length, and from this it follows that *GF* and *GE* are also perpendicular and of equal length.

An immediate consequence of this is that *G* is the midpoint of the square *EXYF* erected on the segment *EF*, as shown in Figure 129b-P. Since Δ*EGF* is an isosceles right triangle, it is one quarter of this square, and *G* is therefore its midpoint.

The argument can be repeated in exactly the same way for points *H* and *J*, and this means that the midpoints *G*, *H*, and *J* of the parallelograms *ARZS*, *BTXM*, and *CNYQ* are simultaneously the midpoints of the squares erected on the sides of triangle *DEF*, as shown

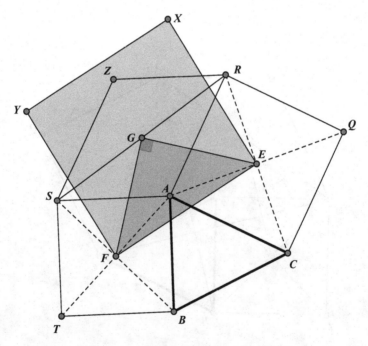

Figure 129b-P

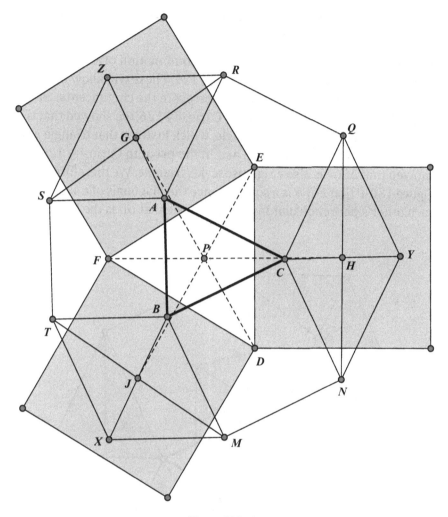

Figure 129c-P

in Figure 129c-P. We can now complete the argument in the same way we did in Part 3 of the proof to Curiosity 126. We see that *DG*, *EJ*, and *FH* are the altitudes of triangle *DEF*, and these are known to intersect in the orthocenter *P* of *DEF*, which is what we had set out to prove.

Curiosity 130. Another Surprising Concurrency from Squares on Triangle Sides

This result can be proved by applying a combination of the results we have established in Curiosities 126 and 129. These will allow us to see that the lines *GU, JV,* and *HW* must intersect in the circumcenter of triangle *DEF*. In Part 2 of the proof for Curiosity 126, we showed that triangle *EFZ* is a right isosceles triangle, which justifies that triangle *EFU* is also a right isosceles triangle, and in the proof to Curiosity 129, we showed that △*EGF* is also a right isosceles triangle. We therefore see in Figure 130-P that *EGFU* is a square. Since the diagonals of a square are each other's perpendicular bisectors, we see that *GU* is the bisector of

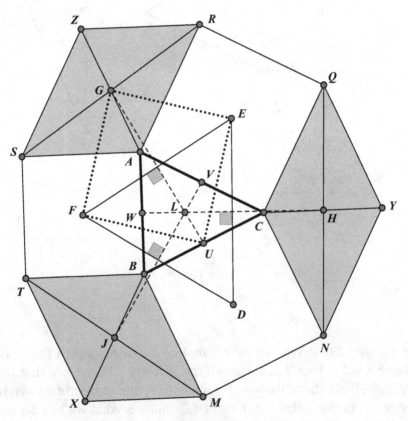

Figure 130-P

EF. Analogously, *JV* is the bisector of *FD* and *HW* is the bisector of *DE*, and the three lines *GU*, *JV*, and *HW* therefore meet in the circumcenter of triangle *DEF*.

Curiosity 131. Using Squares on Triangle Sides to Create another Surprising Concurrency

Let us assume that in Figure 131-P the point *X* is defined as the intersection of lines *ST* and *RQ* and *Y* as the intersection of lines *ST* and *MN*. Furthermore, let *P* denote the intersection of lines *XA* and *YB*. Since *AB∥XY*, triangles *PAB* and *PXY*, which share sides through *P*, are

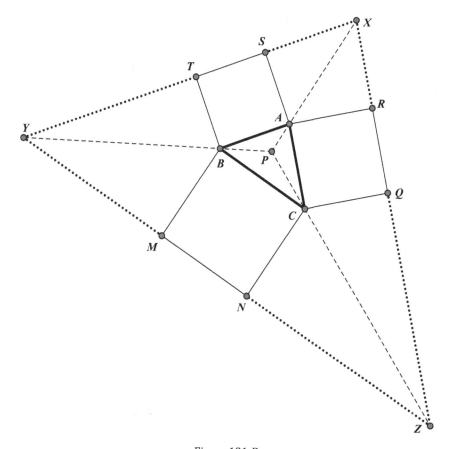

Figure 131-P

similar with $\dfrac{PA}{PX} = \dfrac{PB}{PY} = \dfrac{AB}{XY}$. Next, we define the point Z' on the extension of PC beyond C, such that $\dfrac{PC}{PZ'} = \dfrac{AB}{XY}$. We then have $\dfrac{PA}{PX} = \dfrac{PC}{PZ'}$, and triangles PCA and $PZ'X$, which share sides through P, are similar. This means $XZ' \| AC$, and Z' therefore is on line RQ. Furthermore, triangles PBC and PYZ' are also similar, which yields $YZ' \| BC$; and Z', therefore, is also on line MN. We see that $Z = Z'$, as it is the common intersection point of MN and RQ, and lines AX, BY, and CZ thus all pass through the common point P.

Curiosity 132. Squares on the Legs of a Right Triangle Produce an Unexpected Equality

As is common in right triangles, in Figure 132-P, we let a denote the length of leg BC and b denote the length of leg CA in right triangle ABC, where $\angle ACB = 90°$. Since triangles AJC and FJB are both right triangles and share vertical angles in J, they are similar, and we

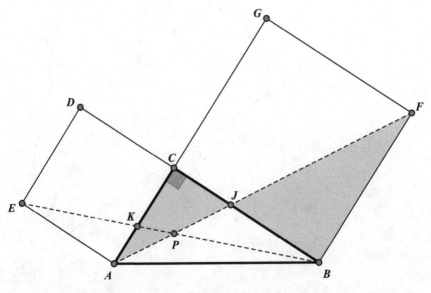

Figure 132-P

have $\dfrac{CJ}{BJ} = \dfrac{AC}{BF} = \dfrac{b}{a}$. Since $BJ = BC - CJ = a - CJ$, this is equivalent to

$\dfrac{CJ}{a-CJ} = \dfrac{b}{a}$, or $a{\cdot}CJ = ab - b{\cdot}CJ$, or $CJ = \dfrac{ab}{a+b}$. Similarly, triangles BCK

and EAK are also similar, and we have $\dfrac{CK}{AK} = \dfrac{BC}{AE} = \dfrac{a}{b}$, from which we

obtain $\dfrac{CK}{b-CK} = \dfrac{a}{b}$, or $b{\cdot}CK = ab - a{\cdot}CK$, or $CK = \dfrac{ab}{a+b}$. We see that

$CJ = CK = \dfrac{ab}{a+b}$, and the two line segments have equal length, as we

had set out to prove.

Curiosity 133. More About Squares on the Legs of a Right Triangle

In Figure 133-P, we first construct altitude HC to hypotenuse AB of right triangle ABC. We can now prove $\triangle AES \cong \triangle CAH$ as follows: $\angle ASE = \angle CHA = 90°$, $AE = AC$, $\angle EAS$ is complementary to $\angle HAC$, and $\angle ACH$ is also complementary to $\angle HAC$. Therefore, $\angle EAS = \angle ACH$. This proves that $\triangle AES \cong \triangle CAH$. In a similar fashion, we can prove that $\triangle FBT \cong$

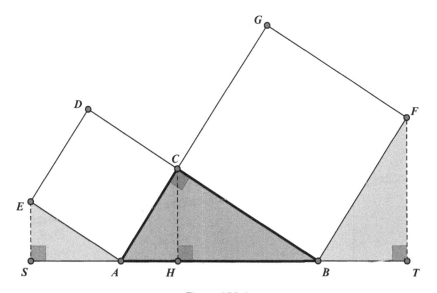

Figure 133-P

Δ*BCH*. From the first congruence, we obtain *ES* = *AH*, and from the second, we obtain *FT* = *BH*. We can then conclude that *ES* + *FT* = *AH* + *BH* = *AB*. Also, from the congruent triangles, we immediately obtain area[*AES*] + area[*FBT*] = area[*CAH*] + area[*BCH*] = area[*ABC*].

Curiosity 134. More Placements of Squares on Right Triangles

As we see in Figure 134-P, we define *X* as the foot of the perpendicular *DX* on the extension of *BC*. Since ∠*BXD* = ∠*ACB*, *AB* = *BD*, and

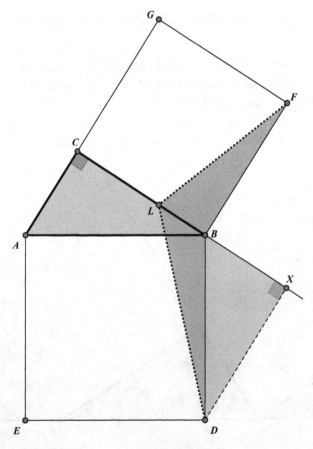

Figure 134-P

$\angle DBX = 180° - \angle ABD - \angle CBA = 180° - \angle CBA = \angle BAC$, we have $\triangle BDX \cong$ $\triangle ABC$. It follows that $DX = BC$; therefore, $\text{area}[LBF] = \frac{1}{2} \cdot LB \cdot BF = \frac{1}{2} \cdot LB \cdot BC = \frac{1}{2} \cdot LB \cdot DX = \text{area}[LDB]$, as we had set out to prove.

Curiosity 135. Another Unexpected Area Equality

As shown in Figure 135-P, we draw point X on AB and point Y on DE such that C, X, and Y lie on a common line parallel to the sides AE and BD of the square $AEDB$. Also, we locate the point P on AE extended beyond A so that $CP \perp AE$ and the point Q on DB extended beyond B so that $CQ \perp BD$. We then have

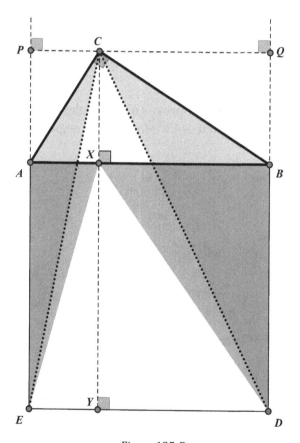

Figure 135-P

$$\text{area}\left[ACE\right]+\text{area}\left[BCD\right]=\frac{1}{2}\cdot AE\cdot PC+\frac{1}{2}\cdot BD\cdot QC$$

$$=\frac{1}{2}\cdot AE\cdot AX+\frac{1}{2}\cdot BD\cdot BX$$

$$=\frac{1}{2}\cdot AE\cdot AX+\frac{1}{2}\cdot AE\cdot XB$$

$$=\frac{1}{2}\cdot AE\cdot\left(AX+XB\right)$$

$$=\frac{1}{2}\cdot AE\cdot AB$$

$$=\frac{1}{2}\text{area}\left[AEDB\right]$$

as we had set out to prove.

Curiosity 136. The Square on the Hypotenuse of a Right Triangle

In Figure 136-P, we will first prove that triangles *ADM* and *BCM* are congruent. We can see that *MA* = *MB*, as both segments are half the

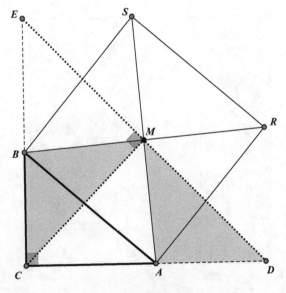

Figure 136-P

length of the diagonals of square *ARSB*. We next note that both ∠*DAR* and ∠*CBA* are complementary to ∠*BAC*. Therefore, ∠*DAR* = ∠*CBA*. It then follows that since ∠*RAM* = ∠*ABM* = 45° we have ∠*DAM* = ∠*CBM*. We also have ∠*AMD* = ∠*BMC* since each of these angles results from removing ∠*CMA* from a right angle, namely, ∠*AMR* and ∠*BMA*, respectively. Therefore, we have △*ADM* ≅ △*BCM*, and thus, *BC* = *AD*. In an analogous fashion, we can prove △*BME* ≅ △*AMC*, which gives us *EB* = *CA*. By adding these two equalities, we have *CD* = *CA* + *AD* = *ED* + *BC* = *EC*, establishing △*CDE* as an isosceles right triangle. Therefore, its altitude *CM* is the bisector of ∠*DCE*.

Curiosity 137. A Truly Unexpected Collinearity

In Figure 137-P, we begin by noticing that the two equilateral triangles *BEC* and *DFC* have the same length base, and so *EC* = *FC*. We then have ∠*FCE* = ∠*FCD* + ∠*DCB* − ∠*ECB* = 60° + 90° − 60° = 90°. which determines that triangle *FEC* is an isosceles right triangle. Also, triangle *ABE* is an isosceles triangle with ∠*EBA* = 30°; therefore, $\angle AEB = \frac{1}{2}(180° - \angle EBA) = \frac{1}{2}(180° - 30°) = 75°$. We know that ∠*BEC* = 60°. Therefore, ∠*AEB* + ∠*BEC* + ∠*CEF* = 75° + 60° + 45° = 180°, and *A*, *E*, and *F* are thus collinear.

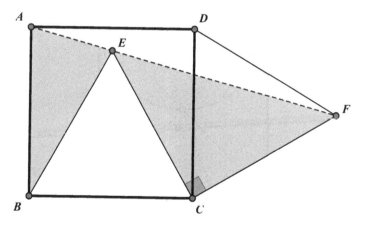

Figure 137-P

Curiosity 138. A Most Unusual Procedure to Divide a Square into Two Equal Parts

As we see in Figure 138-P, we draw the circle with diameter AD. Since $\angle AED$ is a right angle, point E must lie on this circle. Now, let point M denote the second common point of EL with the circle. Since EL is the bisector of the right angle $\angle AED$, we have $\angle MAD = \angle MED = \frac{1}{2}\angle AED = 45°$.

Since $\angle CAD = 45°$, point M lies on the diagonal AC of the square. Since we can repeat this argument on the other side, we also obtain $\angle ADM = \angle AEM = \frac{1}{2}\angle AED = 45° = \angle ADB$, and point M also lies on the diagonal BD. This means that M is the midpoint of the square $ABCD$. Since M lies on EL, the two parts of the square resulting from dividing it with this line, namely, $ABLK$ and $DKLC$, are thus symmetric with respect to M. It follows that they are congruent, and consequently have equal areas.

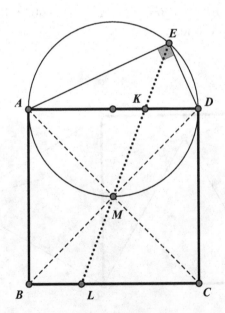

Figure 138-P

Curiosity 139. Doubling a Square

As shown in Figure 139-P, triangle ADB is an isosceles right triangle, since $\angle DAB = 45°$. Therefore, $\dfrac{AD}{AB} = \dfrac{1}{\sqrt{2}}$. Consider quadrilateral $BEDC$, where the two right triangles BDC and BEC share a common hypotenuse BC, so that quadrilateral $BEDC$ is cyclic. Therefore, $\angle DCB + \angle DEB = 180°$ and $\angle AED + \angle DEB = 180°$, so that $\angle DCB = \angle AED$ and then $\triangle AED \sim \triangle ABC$. We then have $\dfrac{DE}{BC} = \dfrac{AD}{AB} = \dfrac{1}{\sqrt{2}}$. Therefore, we have $BC = DE\sqrt{2}$, and thus $\operatorname{area}[BGFC] = BC^2 = \left(DE\sqrt{2}\right)^2 = 2DE^2 = 2 \cdot \operatorname{area}[DHIE]$, which is what we sought to prove.

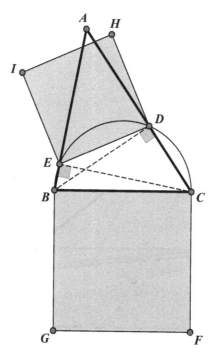

Figure 139-P

Curiosity 140. A Strange Construction of Parallel Lines

We are given the triangle ABC, the point P on side AB, and the median AD, as shown in Figure 140-P. The point G is the common point of AD and PC. Let us assume that we determine a point Q' such that triangles GBC and $GQ'P$ are similar. Since $\angle GCB = \angle GPQ'$, we know that PQ' is parallel to BC. Also, since $\angle BGC = \angle Q'GP$ and P, G, and C lie on a common line, the same is true of points B, G, and Q'. It only remains to be shown that Q' lies on AC, as it will then imply $Q' = Q$.

Let K denote the point in which PQ' intersects with the median AD. Since $\angle DGC = \angle KGP$ and $\angle GCB = \angle GPQ'$, triangles GDC and GKP are similar. Since D is the midpoint of BC, it implies that K is the midpoint of PQ'. However, since K is a point on the median AD, the point symmetric to P with respect to K must lie on triangle side AC. This implies that $Q' = Q$, as required, and we see that PQ is parallel to BC.

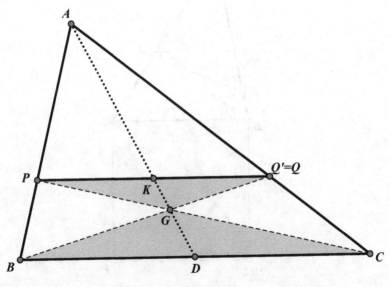

Figure 140-P

Curiosity 141. An Unusual Technique to Find the Midpoint of a Line Segment

The technique we use for this construction uses the same auxiliary lines as in the previous example, as we see in Figure 141-P. However, we should note that the order in which the lines are drawn has an impact on the procedure we will use for our proof. While we once again consider similar triangles, as we did in the previous example, the way we do so is quite different. Let us consider triangles GMC and GNP. These triangles are certainly similar, since PQ and BC are parallel, and we therefore have $\angle GCM = \angle GPN$ and $\angle CMG = \angle PNG$. Similarly, $\triangle GBM$ and $\triangle GQN$ are also similar, since $\angle MBG = \angle NQG$ and $\angle GMB = \angle GNQ$. Furthermore, these triangle pairs have common sides GM and GN, and this gives us $\dfrac{BM}{QN} = \dfrac{MG}{NG} = \dfrac{CM}{PN}$, or just $\dfrac{BM}{QN} = \dfrac{CM}{PN}$. This is equivalent to $\dfrac{BM}{CM} = \dfrac{QN}{PN}$. We now repeat this line of reasoning for different pairs of

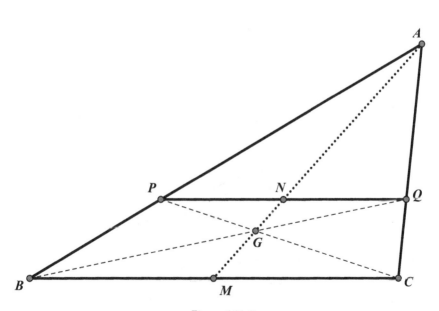

Figure 141-P

similar triangles. Since $\angle MBA = \angle NBA$ and $\angle AMB = \angle ANP$, we see that triangles ABM and APN are similar. Also, since $\angle CMA = \angle QNA$ and $\angle ACM = \angle AQN$, we also see that triangles AMC and ANQ are similar. Since these triangle pairs have common sides AM and AN, this gives us

$$\frac{BM}{PN} = \frac{MA}{NA} = \frac{CM}{QN}, \text{ or just } \frac{BM}{PN} = \frac{CM}{QN}, \text{ and this is equivalent to } \frac{BM}{CM} = \frac{PN}{QN}.$$

Since this implies $\dfrac{QN}{PN} = \dfrac{PN}{QN}$, we have $PN = NQ$, and similarly also $BM = MC$, as was to be proved.

Curiosity 142. The Unexpected Appearance of Parallel Lines

We begin by recognizing that D, E, and F are the midpoints of sides BC, CA, and AB, respectively, of triangle ABC, as shown in Figure 142-P. We also have $GD \| AB$ and $HD \| AC$. Since triangles EGA and FAH both

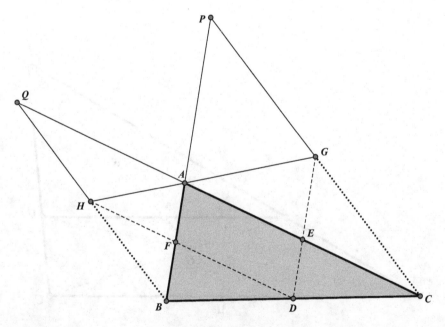

Figure 142-P

have sides on *GH*, and the previously mentioned parallels determine that ∠*AHF* = ∠*GAE* and ∠*HAF* = ∠*AGE* so that triangles *EGA* and *FAH* are similar, we then obtain $\dfrac{GE}{AE} = \dfrac{AF}{HF}$. We draw *P* on the extension of *CG* so that *CP* = 2*CG*, as shown in Figure 142-P. Since we also have *CA* = 2*CE*, triangles *CGE* and *CPA* are similar, with *AP* = 2*EG* and *AP*∥*EG*, which implies that *P* lies on the extension of *BA* beyond *A*. Similarly, drawing *Q* on the extension of *BH* such that *BQ* = 2*BH* yields similar triangles *BFH* and *BAQ*, with *AQ* = 2*FH* and *Q* on the extension of *CA* beyond *A*. From $\dfrac{GE}{AE} = \dfrac{AF}{HF}$, we therefore obtain the following: $\dfrac{AP}{AC} = \dfrac{2GE}{2AE} = \dfrac{2AF}{2HF} = \dfrac{AB}{AQ}$. Since triangles *APC* and *AQB* have equal vertical angles at point *A*, namely, ∠*BAQ* = ∠*PAC*, the triangles are similar with ∠*ABQ* = ∠*APC*. Since this is equivalent to ∠*PBQ* = ∠*BPC*, we thus obtain *PC*∥*QB*, or *GC*∥*HB*.

Curiosity 143. The Unanticipated Parallel Line

We begin the proof by extending *AD* and *AE* to meet *BC* at points *F* and *G*, respectively, in Figure 143-P. Since *BD* is perpendicular to *ADF*, we

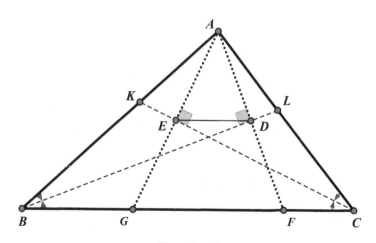

Figure 143-P

see that $\triangle ADB \cong \triangle FBD$. Therefore, $AD = DF$. Analogously, since $\triangle ACE \cong \triangle GEC$, we also have $AE = EG$. We notice that points D and E are the midpoints of AF and AG, respectively. Therefore, DE is parallel to FG, so that DE is parallel to BC.

Curiosity 144. The Surprising Perpendicularity

We begin the proof by applying Menelaus' theorem (see Toolbox) to the intersections of EG with the three sides of triangle ABC, as we see in Figure 144-P. From this, we get $\dfrac{AE}{BE} \cdot \dfrac{BG}{CG} \cdot \dfrac{CF}{AF} = 1$. Since DE and DF are angle bisectors, we have $\dfrac{AE}{BE} = \dfrac{AD}{BD}$ and $\dfrac{CF}{AF} = \dfrac{CD}{AD}$, respectively (see Toolbox). Then, by multiplying these two equations, we get $\dfrac{AE}{BE} \cdot \dfrac{CF}{AF} = \dfrac{AD}{BD} \cdot \dfrac{CD}{AD} = \dfrac{CD}{BD}$. Similarly, since AD bisects $\angle BAC$, we have $\dfrac{CD}{BD} = \dfrac{AC}{AB}$. From Menelaus' theorem, we therefore get $\left(\dfrac{AE}{BE} \cdot \dfrac{CF}{AF}\right) \cdot \dfrac{BG}{CG} = \dfrac{CD}{BD} \cdot \dfrac{BG}{CG} = \dfrac{AC}{AB} \cdot \dfrac{BG}{CG} = 1$, or $\dfrac{BG}{CG} = \dfrac{AB}{AC}$. Thus, with H denoting a point on EA extended, AG is a bisector of $\angle CAH$, which is the exterior angle of triangle ABC. Since we know that the bisectors of

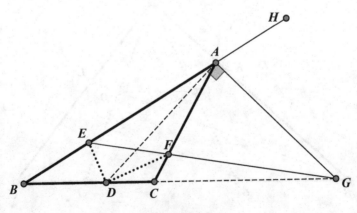

Figure 144-P

two adjacent supplementary angles ($\angle ABC$ and $\angle GAC$) are perpendicular, we have $AG \perp AD$.

Curiosity 145. Another Unexpected Perpendicularity

Since right angles $\angle APB$ and $\angle AHB$ share the same hypotenuse AB, they create a cyclic quadrilateral $ABHP$, as shown in Figure 145-P. Since the opposite angles of a cyclic quadrilateral are supplementary, $\angle APH + \angle CBA = 180°$. Since $\angle QPA$ is also supplementary to $\angle APH$, we have $\angle CBA = \angle QPA$. Furthermore, quadrilateral $AHCQ$ is also a cyclic quadrilateral since the opposite angles at vertices H and Q are supplementary, and we then have $\angle AQH = \angle ACB$. Thus, $\triangle APQ \sim \triangle ABC$. Since AR is a median of $\triangle APQ$, and AM is the corresponding median of $\triangle ABC$, we have $\angle ARH = \angle AMH$. Once again, we have a cyclic quadrilateral $AHMR$, since AH subtends equal angles at points R and M. The opposite angles of a cyclic quadrilateral are supplementary, and $\angle MHA$ is a right angle, so it must follow that $\angle ARM$ is also a right angle.

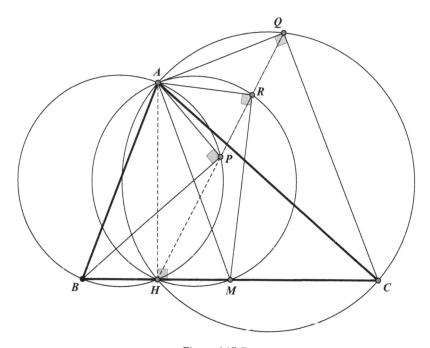

Figure 145-P

Curiosity 146. Yet Another Unexpected Right Angle

In order to prove that angle $\angle HPJ$ is a right angle, we will show that
$NJ = NH = NP$, which will then indicate that the points J, P, and H lie on
a circle with N as its center, thus making $\angle JPH$ a right angle, as shown
in Figure 146-P. In triangle AOC, where $NH \| OC$, we have $\dfrac{AN}{AH} = \dfrac{AO}{AC}$.

Also, in triangle AFO, where $NP \| OF$, we have $\dfrac{AN}{NP} = \dfrac{AO}{OF}$. However,

$OC = OF$, as they are radii of the same circle, and thus $\dfrac{AN}{NP} = \dfrac{AO}{OC}$. Since

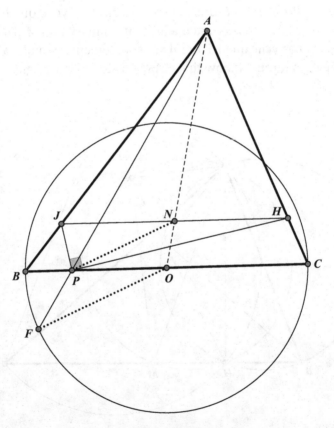

Figure 146-P

$\dfrac{AO}{OC} = \dfrac{AN}{NH}$, we therefore have $\dfrac{AN}{NH} = \dfrac{AN}{NP}$, and thus $NH = NP$. Since AO is a median of triangle ABC, and JH is parallel to BC, then $NJ = NH$. We thus have $NJ = NH = NP$, which is what we set out to prove to establish that angle JPH is inscribed in the semicircle whose diameter is JH, thus making $\angle JPH = 90°$.

Curiosity 147. Four Important Concyclic Points

We begin in Figure 147-P by noting that quadrilateral $ARHQ$ is cyclic, as we have $\angle HRA = \angle AQH = 90°$. We therefore have $\angle QHR = 180° - \angle RAQ$. Since $\angle QHR$ and $\angle BHC$ are vertical angles, they are equal, and we have $\angle BHC = 180° - \angle RAQ = 180° - \angle BAC$.

Since K was defined by $PH = PK$ and $HK \perp BC$, triangles HBC and KCB are symmetric with respect to BC. We then obtain $\angle CKB = \angle BHC = 180° - \angle BAC$, which makes quadrilateral $KBAC$ cyclic. Therefore, the points A, B, K, and C lie on a common circle, as we set out to prove.

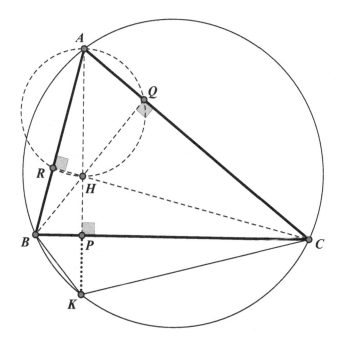

Figure 147-P

Curiosity 148. Four Remarkable Concyclic Points

We begin in Figure 148-P by noticing that $\angle AOB = 2\angle ACB$ since they are both measured by the same intercepted arc AB. In isosceles triangle ABO, we then have $\angle BAO = \frac{1}{2}(180° - \angle AOB) = 90° - \angle ACB$. The altitude from point A intersects BC at point G, and we let H denote the intersection of OA and EF. In right triangle AGC, we have $\angle GAC = 90° - \angle ACB$. Therefore, $\angle BAO = \angle GAC$. Since $\angle AEH = 90° - \angle BAO$ and $\angle ACB = 90° - \angle GAC$, we have $\angle AEH = \angle ACB$. Because $\angle AEH$ is supplementary to $\angle BEF$, $\angle BEF$ is also supplementary to $\angle ACB$. $\angle BEF$

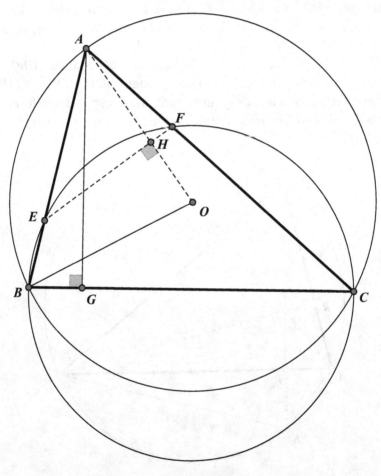

Figure 148-P

and ∠*ACB* are the opposite angles of quadrilateral *BEFC*, and *BEFC* is therefore a cyclic quadrilateral.

Curiosity 149. Four Unexpected Concyclic Points

When we construct the diameter of the circumcircle of triangle *ABC* from point *A* in Figure 149-P, it intersects the circle at point *S*, and ∠*ACS* and ∠*SBA* are both right angles, since they are inscribed in semicircles. Therefore, *BH* is parallel to *CS*, and *BS* is parallel to *CH*, which then makes *BSCH* a parallelogram. Since *HD* and *DS* are portions of the same diagonal *HS*, we have therefore determined that points *X*, *H*, *D*, and *S* are collinear with *HD* = *DS*. Similarly, we can also show the *HF* = *FT*, as they are both extensions of line segments joining the orthocenter *H* to the midpoint of a side of triangle *ABC*.

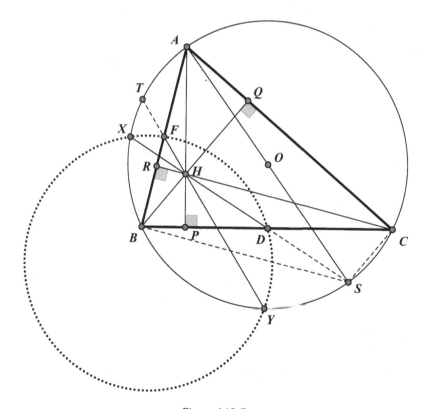

Figure 149-P

Using the well-known relationship that the products of the segments of intersecting chords of a circle are equal, we now have $HY \cdot HT = HS \cdot HX$. Since $2HF = HT$, and $2HD = HS$, then by substitution we have $HY \cdot 2HF = 2HD \cdot HX$, or $HY \cdot HF = HD \cdot HX$. This implies that since the products of the two intersecting segments are equal, their endpoints, Y, D, F, and X, lie on the same circle.

Curiosity 150. Perpendiculars that Generate Concyclic Points

In order to show that quadrilateral $EHJG$ is cyclic, we will show that its opposite angles $\angle EGJ$ and $\angle JHE$ are supplementary, as shown in Figure 150-P. Triangle APQ is a right triangle; therefore, $\angle JPA + \angle AQJ = 90°$. Since $\angle AJP$ and $\angle AGP$ are right angles, the quadrilateral $APGJ$ is cyclic, whereupon it follows that $\angle JPA = \angle JGA$. Similarly, quadrilateral $AJHQ$ is also cyclic, and therefore $\angle AQJ = \angle AHJ$. Since $\angle AGC = 90° = \angle AEC$, we have another cyclic quadrilateral $AGEC$, which yields $\angle EGC = \angle EAC$. Also, since $\angle AEB = 90° = \angle AHB$ we have yet another cyclic quadrilateral $ABEH$, which yields $\angle BHE = \angle BAE$. Finally, since $\angle CGA = \angle AHB = 90°$, we obtain $\angle JGA + \angle CGJ + \angle AHJ + \angle JHB = 180°$. Since $\angle JGA + \angle AHJ = \angle JPA + \angle AQJ = 90°$, we obtain $\angle CGJ + \angle JHB = 90°$, and this allows us to combine these angles to obtain: $\angle EGJ + \angle JHE = \angle EGC + \angle BHE + (\angle CGJ + \angle JHB) = \angle EAC + \angle BAE + (\angle CGJ + \angle JHB) = 90° + 90° = 180°$.

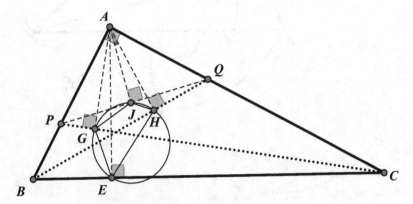

Figure 150-P

We therefore see that the quadrilateral *EHJG* is cyclic, establishing that points *E*, *G*, *H*, and *J* all lie on the same circle.

Curiosity 151. Altitudes and Circles that Generate Another Circle

In order to prove this relationship, we once again apply the fact that the product of the segments of intersecting chords of a circle are equal. In Figure 151-P, let *H* denote the orthocenter of *ABC*, or in other

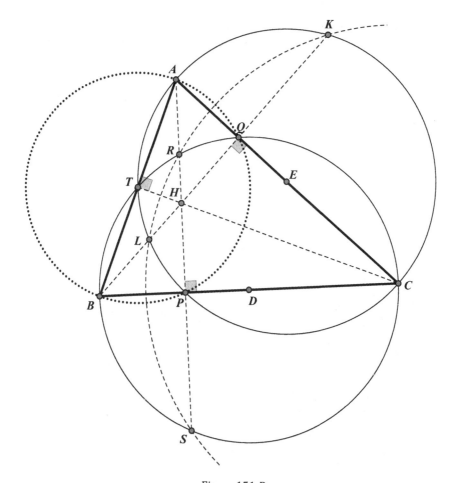

Figure 151-P

words, the common point of altitudes AP and BQ. In the circle with center M, we have $HL \cdot HK = HA \cdot HP$, and in the circle with center N, we have $HR \cdot HS = HB \cdot HQ$. Noting that $\angle AQB = \angle APB = 90°$, we see that P and Q lie on the circle with diameter AB; therefore, $HA \cdot HP = HB \cdot HQ$. From this, we then obtain $HL \cdot HK = HR \cdot HS$, which indicates that the points S, L, R, and K all lie on the same circle. The point T is simply the foot of the altitude from vertex C of triangle ABC to side AB. Since $\angle CTA = \angle BTC = 90°$, this point of AB certainly lies on both circles.

Curiosity 152. More Unexpected Concyclic Points

Since $RD \perp BC$, and $RF \perp AB$, the points B, D, R, and F are concyclic, as we see in Figure 152-P. Therefore, $\angle DBR = \angle DFR$. We extend BP to intersect RF at point X. In right triangle BXF, we have $\angle LFX = \angle FBL$. It then follows that $\angle DFR = \angle XBF$, $\angle CBR = \angle XBA$, and $\angle CBP = \angle RBA$. The perpendicular from point P to BC intersects BC at point I, so that $PI \perp BC$. We then have $BI \cdot BD = (BP \cos \angle CBP) \cdot (BR \cos \angle DBR)$ and $BK \cdot BF = (BP \cos \angle XBF) \cdot (BR \cos \angle RBA)$.

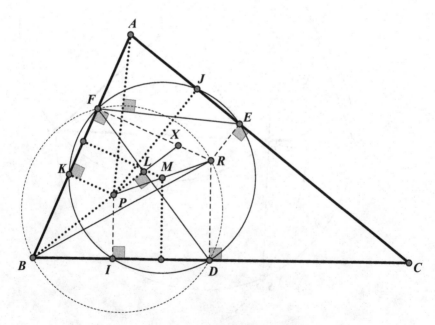

Figure 152-P

It then follows that $BI \cdot BD = BK \cdot BF$; therefore, points K, F, D, and I are concyclic by the tangent-secant theorem. The center M of this circle lies both on the mid-parallel of ID and on the mid-parallel of FK, and is thus the mid-point of PR. If we then have point J on AC such that $PK \perp AC$, we can argue in the same way that the points E, J, F, and K all lie on a common circle with center M. Summing up, we have proved that the points D, E, J, F, K, and I all lie on the same circle, which adds points I and J to our previously concyclic points.

Curiosity 153. A Surprising Five-Point Circle

Since $\angle HPM = 90°$. and an angle inscribed in a semicircle is a right angle, point P lies on the circle with diameter MH. We will show in Figure 153-P that the same is true at points E and F. Since M is the midpoint of BC, and E is the midpoint of BQ, we have that ME is parallel to CQ. Since $\angle BQC = 90°$, we therefore also have $\angle BEM = 90°$, and E, then, also lies on the circle with diameter MH. The same holds for F on the altitude CR, and we see that P, F, and E all lie on the circle with diameter MH.

It remains to be shown that triangles PFE and ABC are similar. Since P, F, H, and E lie on a common circle, and PCH and BCR are right triangles, we have $\angle PEF = \angle PHF = \angle PHC = 90° - \angle HCP = 90° - \angle RCB = \angle CBA$.

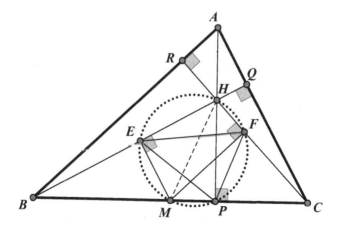

Figure 153-P

Similarly, we also obtain $\angle EFP = \angle ACB$; therefore, triangles *PFE* and *ABC* are similar.

Curiosity 154. The Famous Nine-Point Circle

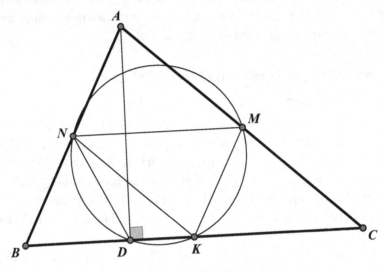

Figure 154a-P

In Figure 154a-P, points *K*, *M*, and *N* are the midpoints of the three sides of triangle *ABC*, and *AD* is an altitude of triangle *ABC*. Since *MN* is a midline of triangle *ABC*, we have *BC*∥*MN*. Therefore, *DKMN* is a trapezoid. Furthermore, *KM* is also a midline of triangle *ABC*, so that $KM = \frac{1}{2}AB$. Since *ND* is the median to the hypotenuse of right triangle *BDA*, we also have $ND = \frac{1}{2}AB$. Therefore, *KM* = *ND* and trapezoid *DKMN* is isosceles. We recall that when the opposite angles of a quadrilateral are supplementary, as in the case of an isosceles trapezoid, the quadrilateral is cyclic. Therefore, quadrilateral *DKMN* is cyclic, and *D* lies on the circumcircle of triangle *KMN*.

Using the same procedure, we also find that quadrilaterals *ENKM* and *FNKM* are also isosceles trapezoids and therefore cyclic. From this, we see that *E* and *F* also lie on the circumcircle of triangle *KMN*.

We now turn our attention to points *X*, *Y*, and *Z*.

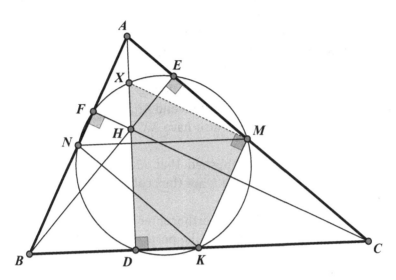

Figure 154b-P

With *H* as the orthocenter (the point of intersection of the altitudes) of triangle *ABC*, *X* is the midpoint of *AH*, as we see in Figure 154b-P. Therefore, *MX*, which is a midline of triangle *AFC*, is parallel to the altitude *CF* of *ABC*. Furthermore, since *KM* is a midline of triangle *ABC*, we have *KM*∥*AB*. Therefore, since ∠*CFA* is a right angle, we know that ∠*XMK* = ∠*CFA* = 90°. Since *KD* ⊥ *AD*, quadrilateral *DKMX* is cyclic (recall that the opposite angles are supplementary). This places point *X* on the circumcircle of *DKM*, which we have already identified as the circumcircle of triangle *KMN*. With analogous arguments, we also place *Y* and *Z* on the circumcircle of triangle *KMN*, and we see that all nine points, *K*, *M*, *N*, *D*, *E*, *F*, *X*, *Y*, and *Z*, lie on this circle, known as the famous *Nine-Point Circle*.

Curiosity 155. A Collinearity with the Center of the Nine-Point Circle

We begin in Figure 155-P by recalling that *XK* is a diameter of the Nine-Point Circle and point *R* is the midpoint of *XK*. This results from the fact that *XNKZ* is a rectangle, and the diagonal *XK* is a diameter of its circumcircle, which also happens to be the Nine-Point Circle. To show this, we note that *N* and *K* are the midpoints of *BA* and *BC*, respectively, which means that *NK* is parallel to *AC* and *NK* is half as long as *AC*. Similarly, since *Z* and *X* are the midpoints of *HC* and *HA*, respectively, *ZX* is also parallel to *AC* and also half as long as *AC*. This results in recognizing *XNKZ* as a parallelogram. In addition to this, since *N* and *X* are the midpoints of *AB* and *AH*, respectively, *NX* is parallel to *BH*. Since *BH* ⊥ *AC*, we also have *NX* ⊥ *ZX*, and parallelogram *XNKZ* is therefore a rectangle.

Now that we have established that *XK* is a diameter of the Nine-Point Circle and *X*, *R*, and *K* are thus collinear, we wish to show that *S* lies on this diameter.

From the definition of the orthocenter *H* in triangle *ABC* and the fact that point *S* is the foot of the perpendicular from the orthocenter *H* to the bisector *AT* of ∠*BAC*, we have *FH* ⊥ *FA*, *EH* ⊥ *EA*, and *SH* ⊥ *SA*. This means that points *F*, *E*, and *S* all lie on the circle with diameter *AH*. Since *X* is the midpoint of *AH*, it is the center of this circle with radius *XH*. Since *AT* is the angle bisector of ∠*BAC*, we have

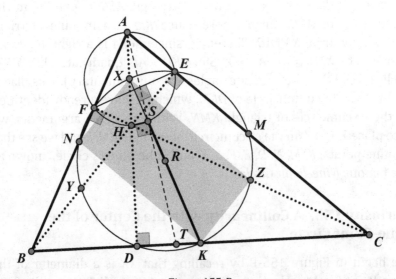

Figure 155-P

$\angle FAS = \angle BAT = \angle TAC = \angle SAE$; thus, $SF = SE$. This means that $XFSE$ is a deltoid (or kite), since we have both $XF = XE$ and $SF = SE$. Therefore, its diagonals are perpendicular, giving us $XS \perp FE$. Since E and F are both points of the Nine-Point Circle, we also have $RF = RE$. $XFRE$ is therefore also a deltoid (or kite), since we have both $XF = XE$ and $RF = RE$. Its diagonals are also perpendicular, giving us $XR \perp FE$, which means that X, S, and R are collinear. Since we already know that X, R, and K are collinear, this Is also true for S, R, and K, which is what we had originally set out to prove.

Curiosity 156. The Meeting of the Three Famous Triangle Centers

To prove the relationship of the orthocenter, the circumcenter, and the centroid, we will consider two pairs of similar triangles in Figure 156-P. First, consider triangles ABH and DEO. Since D is the midpoint of BC and E the midpoint of AC, we know that DE is parallel to AB and also half as long as AB. Furthermore, since both AH and DO are perpendicular to BC, they are parallel. Likewise, since both BQ and EO are perpendicular to AC, they are parallel. We see that each pair of sides of triangles ABH and DEO are parallel; therefore, the triangles are similar. Since we already know that AB is twice the length of ED, AH is also twice the length of OD.

We define G as the point of intersection of the line OH and the median AD, and then we can consider the triangles AHG and DOG. We have already established that AH and DO are parallel, and since AG and DG lie on the same line, as do HG and OG, triangles AHG and DOG are also similar. Because we know that AH is twice the length of OD, we now know that AG is also twice the length of DG. This means that the point G on the median AD is the centroid of ABC, as it divides the median in the ratio 2:1. We see that the centroid of triangle ABC does indeed lie on OH, as claimed, and because of the similarity of the triangles AHG and DOG, we also have $OG = \frac{1}{2}HG$. The remarkable line containing the three points O, G and H is commonly referred to as the *Euler Line* of triangle ABC, named after the famous Swiss mathematician Leonhard Euler (1707–1783).

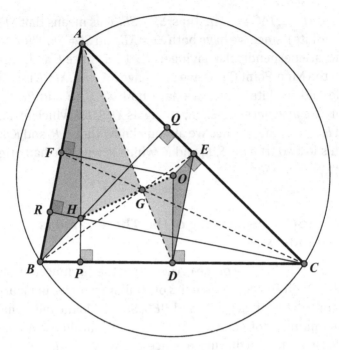

Figure 156-P

Curiosity 157. Properties of the Nine-Point Circle

We first note that the circumcenter O lies on the perpendicular bisector of BC and the center of the Nine-Point Circle R lies on the perpendicular bisector of DK. As we see in Figure 157-P, this means that both of these perpendicular lines are parallel to the altitude AD of triangle ABC. We see that quadrilateral $DKOH$ is a trapezoid with two consecutive right angles and the perpendicular bisector of DK as its mid-line. This mid-line intersects the side OH in its midpoint. Since an analogous argument holds for the trapezoids $MEHO$ and $FNOH$, we see that the center R of the Nine-Point Circle is the midpoint of OH, as we had set out to show.

Having established this, it is now quite easy to see why the radius of the Nine-Point Circle is half that of the circumcircle. Since Y is the midpoint of BH, and R is the midpoint of HO, we have $\dfrac{HY}{HB} = \dfrac{HR}{HO} = \dfrac{1}{2}$.

This means that the triangles HYR and HBO are similar, with $\dfrac{RY}{OB} = \dfrac{1}{2}$

or $OB = 2RY$. Since OB is a radius of the circumcircle and RY is a radius of the Nine-Point Circle, the proof is therefore complete.

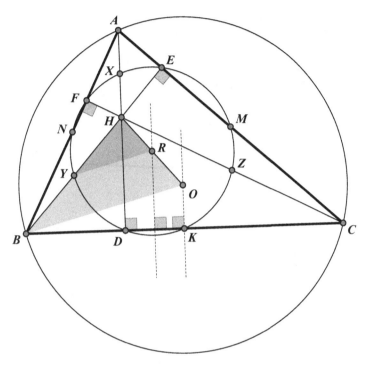

Figure 157-P

Curiosity 158. More Properties of the Nine-Point Circle

This result is an immediate consequence of Curiosity 157. There, we established that $\dfrac{HR}{HO}=\dfrac{1}{2}$ and $\dfrac{RL}{OJ}=\dfrac{1}{2}$ hold. Since triangles HRL and HOJ, shown in Figure 158-P, have a common angle in H, they are therefore similar. We, thus, have $\dfrac{HL}{HJ}=\dfrac{1}{2}$, as we set out to show.

Curiosity 159. An Unexpected Collinearity

Unexpectedly, this property is an immediate consequence of the existence of the Nine-Point Circle, which we first encountered in

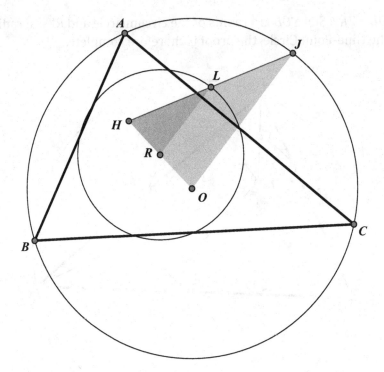

Figure 158-P

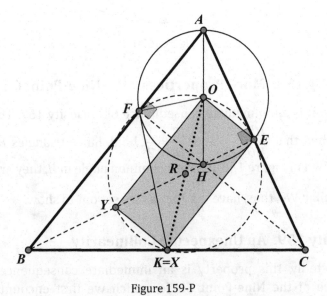

Figure 159-P

Curiosity 154. Recall that this circle contains the feet of the altitudes, the midpoints of the sides of the triangle, and the midpoints of the three segments joining the orthocenter H to the vertices of the triangle.

In Figure 159-P, we see the Nine-Point Circle with center R. The quadrilateral $OEKY$ is shaded, and we note that the vertices of this quadrilateral all lie on the Nine-Point Circle. Points E and K are the midpoints of sides CA and BC, respectively, while Y is the midpoint of HB. We already know that O is the center of the circle through A and H, and since AH must be a diameter of this circle, O is the midpoint of AH. Therefore, O is also on the Nine-Point Circle. Since O and Y are the midpoints of HA and HB, respectively, triangles HAB and HOY are similar, and we have $OH = \frac{1}{2} \cdot AB$ and $OY \| AB$. Also, since E and K are the midpoints of CA and BC, respectively, triangles CAB and CEK are also similar, and we have $EK = \frac{1}{2} \cdot AB$ and $EK \| AB$. This means that since OY and EK are of equal length and parallel, $OEKY$ is a parallelogram. In addition, we note that O and E are the midpoints of AH and AC, respectively, of triangle AHC so that triangles AHC and AOE are similar. Then $OE = \frac{1}{2} HC$, and $OE \| HC$. Since the altitude HC of triangle ABC is perpendicular to side AB, we then have OE perpendicular to OY, and the parallelogram $OEKY$ is also a rectangle.

We see that the radius OE of circle O is perpendicular to EK, thus, establishing that EK is tangent to circle O at point E. Analogously, we have FK as the tangent of circle O at point F. This means that the tangents to circle O at points E and F intersect in the midpoint K of triangle side BC.

In summary, we not only see that these two tangents intersect in a point X on the side BC, but as an extra added attraction, they will intersect in the midpoint $K = X$ of BC.

Curiosity 160. A Concurrency Generated by the Orthic Triangle

To prove this concurrency, we recall that the Nine-Point Circle of a triangle ABC is the common circumcircle of the triangles DEF and MNR, as we see in Figure 160-P. We will show that lines JM, KN, and LR meet at the center, *P*, of this circle. Consider the triangle *EFM*. The center, *P*, of the Nine-Point Circle lies on the bisector of its side *EF*. If we can show that triangle *EFM* is isosceles with *ME* = *MF*, then this bisector is also the altitude *MJ* of triangle *EFM*. We already know from Curiosity 20 that $\angle AEF = \angle CBA = \beta$. Furthermore, in the Nine-Point Circle, we have $\angle MEN = \angle MRN = \angle ACB = \gamma$ since the sides of triangle *MNR* are parallel to the sides of triangle *ABC*. This gives us $\angle FEM = 180° - \angle AEF - \angle MEN = 180° - \beta - \gamma = \angle BAC = \alpha$. Repeating this for the other side gives us $\angle EFA = \angle ACB = \gamma$, $\angle RFM = \angle RNM = \angle CBA = \beta$, and $\angle MFE = 180° - \angle EFA - \angle RFM = 180° - \gamma - \beta = \angle BAC = \alpha$, and we see that triangle *EFM* is isosceles with *ME* = *MF*, as we anticipated. We see that the center, *P*, of the Nine-Point Circle lies on *MJ*, and since we can repeat this argument for both *NK* and *RL*, we see that the three lines do indeed have a common point, namely, *P*.

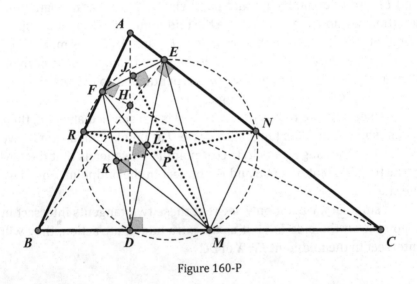

Figure 160-P

Curiosity 161. Altitudes Produce a Concurrency and Equality

From Curiosity 154, we know that the midpoints *K*, *M*, and *N* of the sides of triangle *ABC*, the feet *D*, *E*, and *F* of the altitudes of triangle

ABC, and the midpoints *T*, *U*, and *V* of the segments joining the vertices of triangle *ABC* to its orthocenter *H* all lie on the Nine-Point Circle of triangle *ABC*. We need to show that *TK*, *UM*, and *VN* are diameters of this Nine Point Circle, as they are then certainly concurrent at their midpoints and also of equal length.

In Figure 161-P, we see the quadrilateral *NUVM*. Since *N* and *M* are the midpoints of *AB* and *AC*, respectively, *NM* is parallel to *BC* and half as long as *BC*. Similarly, since *U* and *V* are the midpoints of *HB* and *HC*, the same is true of *UV*. We have therefore established that *NUVM* is a parallelogram. Noting that *U* and *N* are the mid-points of *BA* and *BH*, respectively, it follows that *UN* is also parallel to *HA*. Since *HA* ⊥ *BC*, we thus have *UN* ⊥ *UV*, and parallelogram *NUVM* is then a rectangle.

The center of the circumcircle of this rectangle is the intersection of its diagonals *UM* and *VN*, and these two segments are diameters of the circumscribed circle of rectangle *NUVM*. In analogous fashion for the quadrilaterals *TUKM* and *TNKV*, we also obtain *TK* as a diameter of the circle, and we see that the center *P* of the Nine-Point Circle is a common midpoint of all three segments *TK*, *UM*, and *VN*, which are also of equal length.

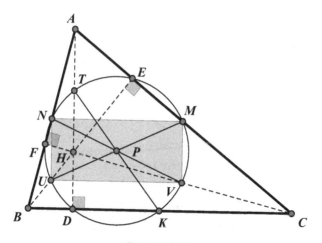

Figure 161-P

Curiosity 162. Napoleon's Contribution to Mathematics

This can easily be justified by showing that the two triangles ABE and DBC are congruent, both of which are shown shaded in Figure 162-P. The fact that the triangles ADB and BEC are equilateral gives us $BA = BD$, $BE = BC$, and $\angle CBD = \angle CBA + \angle ABD = \angle CBA + 60° = \angle CBA + \angle EBC = \angle EBA$.

Therefore, triangles ABE and DBC are congruent. From this, we immediately obtain $AE = CD$. Similarly, we can also show in an analogous way that triangles ABF and ADC are congruent, which then gives us the equality $AE = BF = CD$. This could also have been done using $\triangle CFB$ and $\triangle CAE$.

In order to show that these three lines have a common point (are concurrent), we will have to take a slightly closer look at the situation.

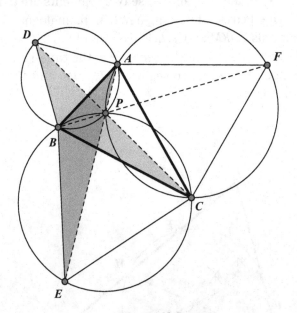

Figure 162-P

Letting P denote the point in which AE and CD intersect, we note that $\angle BAP = \angle BAE = \angle BDC = \angle BDP$ because the triangles ABE and DBC are congruent. This means that points P, A, D, and B lie on a common circle, and we therefore have $\angle APB = 180° - \angle BDA = 180° - 60° = 120°$. Similarly, we also have $\angle PCB = \angle DCB = \angle AEB = \angle PEB$, which means that points P, B, E, and C also lie on a common circle, and we then also have $\angle BPC = 180° - \angle CEB = 180° - 60° = 120°$. It thus follows that $\angle CPA = 360° - \angle APB - \angle BPC = 360° - 120° - 120° = 120°$, and since $\angle AEC = 60°$, points A, P, C, and F also lie on a common circle. This means that $\angle FPA = \angle FCA = 60°$, from which we obtain $\angle APB + \angle FPA = 120° + 60° = 180°$. The point P therefore also lies on BF, and we then have P as a common point of AE, BF, and CD.

Curiosity 163. Napoleon's Minimum Distance Point (The Fermat Point)

We have already shown $\angle APB = \angle BPC = \angle CPA = 120°$ as part of the proof to Curiosity 162. What remains to be shown is that the sum of the distances from P to the three triangle vertices, $PA + PB + PC$, is minimal among all point in the interior of the triangle. To show this, we choose any point Q, not necessarily in the interior of the triangle, as shown in Figure 163-P.

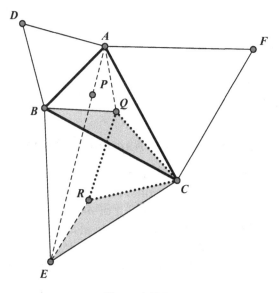

Figure 163-P

We wish to discover a property of the sum of the distances from Q to the triangle vertices, or in other words, the expression $QA + QB + QC$. To facilitate the proof, we add an equilateral triangle CQR as shown in Figure 163-P. Since the sides of this triangle are of equal length, we have $QC = QR$. Also, we note that triangles QBC and REC are congruent, since we have $QC = RC$, $BC = EC$, and $\angle QCB = \angle QCR - \angle BCR = 60° - \angle BCR = \angle BCE - \angle BCR = \angle RCE$. This implies $QB = RE$. We see that $QA + QB + QC = AQ + QR + RE$. If Q does not lie on the line AE, the assumption that this sum is minimal is obviously not correct, since a choice of Q on AE results in a smaller value of the sum. We see that the point P for which the sum $PA + PB + PC$ is minimal must certainly lie on AE. However, we can argue analogously in exactly the same way for BF and CD, and we see that the common point P of AE, BF, and CD we discovered in Curiosity 162 has the property of being the minimal distance point, as we had set out to prove.

Curiosity 164. When the Minimum Distance Point is not Inside the Triangle

As we can see in Figures 164a-P and 164b-P, the property described in Curiosity 163 does not depend on the location of the chosen point Q or the sizes of the interior angles of $\triangle ABC$. We see that the common point P of AE, BF, and CD is always the minimal distance point. When the angle in A is equal to 120°, we have $A = P$ since we always have $\angle DPE = \angle EPF = \angle FPD = 120°$. If the angle at vertex A is greater than 120°, point P will lie outside of $\triangle ABC$.

Curiosity 165. Extensions of Napoleon's Theorem

To prove that triangle KLM is equilateral using Figure 165-P we will show that the sides are proportional to the equal lengths AE, BF, and CD. We first let N denote the midpoint of AC. Since MA bisects $\angle CAF$, we know that $\angle NAM = \angle CAM = 30°$. We also know that point M is the center of triangle ACF; therefore, $MN \perp AC$, which enables us to identify triangle ANM as a 30°-60°-90° triangle. As a result, in triangle ANM, we find that $\dfrac{AN}{AM} = \dfrac{\sqrt{3}}{2}$, or $\dfrac{AC}{AM} = \dfrac{2\sqrt{3}}{2} = \sqrt{3}$. Analogously, in

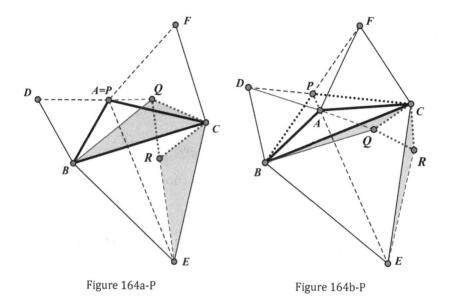

Figure 164a-P   Figure 164b-P

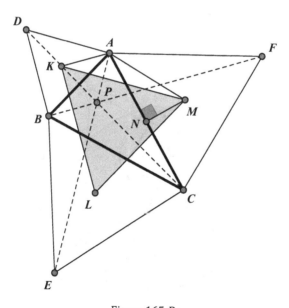

Figure 165-P

triangle *DBA*, we have $\dfrac{AB}{AK} = \sqrt{3}$. We also know that $\angle KAM = \angle KAB +$ $\angle BAM = 30° + \angle BAM = \angle BAM + \angle MAF = \angle BAF$, which then allows us to establish that triangles *AKM* and *ABF* are similar. It then follows that $\dfrac{BF}{KM} = \dfrac{AB}{AK} = \sqrt{3}$. Furthermore, in the same fashion, we can show that $\dfrac{DC}{KL} = \sqrt{3}$, and $\dfrac{AE}{ML} = \sqrt{3}$. We now have $\dfrac{BF}{KM} = \dfrac{DC}{LK} = \dfrac{AE}{ML}$. However, since we know that *BF = DC = AE*, we can conclude that *KM = LK = ML*; therefore, triangle *KLM* is equilateral.

Curiosity 166. Overlapping Side-Equilateral Triangles

Essentially, the proof in this case is the same as the one we have just seen for external equilateral triangles in Curiosity 165, but there are some small details that make the proof worth considering separately. First of all, we can show that lines *AE*, *BF*, and *CD* intersect in a common point *P* with *AE = BF = CD*. In Figure 166-P, we let *P* denote the intersection of lines *AE* and *BF*. As was the case in Curiosity 162, this point *P* is a common point of the circumcircles of the three equilateral

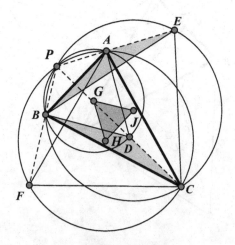

Figure 166a-P

triangles *ABD*, *BCE*, and *CAF*, which we can once again show in exactly the same way as we did in the proof for Curiosity 162, with the aid of congruent triangles *ABE* and *DBC*. It then follows that *P* lies on the line *CD*, as both *PD* and *PC* lie on the angle bisector of ∠*FPE*. This is the case for ∠*FPE* = 120°, as we have ∠*CPE* = ∠*CBE* = 60° and ∠*DPA* = ∠*DBA* = 60°. Also, the congruent triangles *ABE*, *BDC*, and *BAF* give us *AE* = *BF* = *CD* in the same way we were able to show this in the proof to Curiosity 162.

We will now show that triangles *AGH* and *ABF* are similar. We see these triangles in Figure 166b-P, and since $\dfrac{BF}{GH} = \dfrac{AB}{AG} = \sqrt{3}$, we can argue in the same way we did in the proof for Curiosity 165.

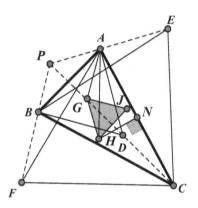

Figure 166b-P

With $\dfrac{DC}{HJ} = \sqrt{3}$, and $\dfrac{AE}{JG} = \sqrt{3}$, we have $\dfrac{BF}{GH} = \dfrac{DC}{HJ} = \dfrac{AE}{JG}$. Since *AE* = *BF* = *CD*, it follows that *JG* = *GH* = *HJ*, and the triangle *GHJ* is therefore equilateral.

Curiosity 167. Surprising Triangle Area Relationship

This unexpected result is not only surprising in itself, but requires a clever proof, which follows a relatively simple application of the law of cosines (see Toolbox).

First, we consider triangle *AKM*, shown in Figure 167-P. Since

$$\angle KAM = \angle KAB + \angle BAC + \angle CAM = 30° + \angle BAC + 30° = \angle BAC + 60°,$$

the law of cosines applied to triangle *AKM* gives us

$$KM^2 = AK^2 + AM^2 - 2AK \cdot AM \cos(\angle BAC + 60°).$$

Similarly, in triangle *AGH*, we have

$$\angle GAH = \angle BAC - \angle BAG - \angle HAC = \angle BAC - 30° - 30° = \angle BAC - 60°.$$

By applying the law of cosines to triangle *AGH*, we get

$$GH^2 = AG^2 + AH^2 - 2AG \cdot AH \cdot \cos(\angle BAC - 60°).$$

Noting that *AG = AK* (as *G* and *K* are symmetric with respect to *AB*), and also *AH = AM* (as *H* and *M* are symmetric with respect to *AC*), we therefore have

$$KM^2 - GH^2 = 2AK \cdot AM \cdot (\cos(\angle BAC - 60°) - \cos(\angle BAC + 60°)).$$

In the proof of Curiosity 165, we have already established that $AK = \dfrac{AB}{\sqrt{3}}$ and $AM = \dfrac{AC}{\sqrt{3}}$. Applying the trigonometric identity $\cos(\alpha - \beta) - \cos(\alpha + \beta) = 2\sin\alpha \sin\beta$ (see Toolbox), we obtain

$$KM^2 - GH^2 = 2 \cdot \frac{AB}{\sqrt{3}} \cdot \frac{AC}{\sqrt{3}} \cdot 2 \cdot \sin\angle BAC \cdot \sin 60° = \frac{2\sqrt{3}}{3} \cdot AB \cdot AC \cdot \sin\angle BAC,$$

which is equivalent to

$$\frac{\sqrt{3}}{4}KM^2 - \frac{\sqrt{3}}{4}GH^2 = \frac{1}{2} \cdot AB \cdot AC \cdot \sin\angle BAC.$$

Recalling that the area of an equilateral triangle with sides of length *a* is equal to $\dfrac{\sqrt{3}}{4}a^2$ and the area of a triangle *ABC* is equal to

$\frac{1}{2}AB \cdot AC \cdot \sin \angle BAC$, this is equivalent to area[KLM] − area[GHJ] = area[ABC], which is what we set out to prove.

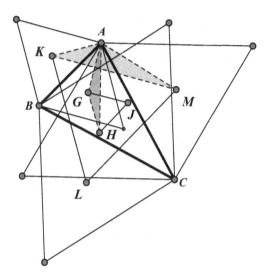

Figure 167-P

Curiosity 168. The Centroid Enters the Previous Configuration

As we see in Figure 168-P, we first draw the perpendicular bisector of BC through L, bisecting BC in X. We then extend LX to L' so that $XL = XL'$. Furthermore, we let Y denote the point in which $L'A$ and KM intersect. Triangle $LL'B$ is equilateral since $LB = L'B$ and $\angle LBL' = 2\angle LBX = 2 \cdot 30° = 60°$. Also, triangles KBL and $ML'L$ are congruent, since we have $BL = LL'$, $KL = ML$, and $\angle KLB = \angle L'LB − \angle L'LK = 60° − \angle L'LK = \angle MLK − \angle L'LK = \angle MLL'$. This gives us $AK = KB = ML'$. Similarly, we can also show that $AM = KL'$ by considering the triangles MLC and KLL'.

This implies that $AKL'M$ is a parallelogram, and since Y is the common point of its diagonals, it is the midpoint of $L'A$. Since X is also the midpoint of $L'L$, we see that AX and LY are medians in triangle $L'AL$, and they intersect in the centroid S of $L'AL$. As centroid in this triangle, point *S* divides both *AX* and *LY* in the ratio 2:1. Since *AX* is also a median of △*ABC*, this means that point *S* is also the centroid of △*ABC*. Furthermore, since *LY* is a median of △*KLM*, point *S* is also the centroid of △*KLM*, and therefore the equilateral triangle's midpoint. We see that the centroid of △*ABC* does indeed coincide with the midpoint of △*KLM*, as we set out to prove.

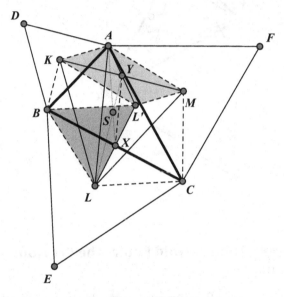

Figure 168-P

Curiosity 169. The Emergence of Another Equilateral Triangle

Since *ADCR* is a parallelogram, we have *AR* = *CD*. In Curiosity 162, we have already established *AE* = *CD*, and this gives us *AE* = *CD* = *AR*. It remains to show that *AE* = *ER* also holds true. In order to demonstrate this, we will show that triangles *CER* and *BEA* are congruent.

Considering Figure 169-P, it is clear that *CR* = *AD* = *AB* and *CE* = *BE* because of the properties of parallelogram *DCRA* and the equilateral triangle *BEC*. Also, since *AC* is a diagonal of parallelogram *DCRA*, we have the following:

$$\angle ECR = 360° - \angle BCE - \angle ACB - \angle RCA$$
$$\angle ECR = 360° - 60° - \angle ACB - \angle DAC$$
$$\angle ECR = 300° - \angle ACB - (\angle DAB + \angle BAC)$$
$$\angle ECR = 300° - (\angle ACB + \angle DAB) - 60°$$
$$\angle ECR = 240° - (180° - \angle CBA)$$
$$\angle ECR = 60° + \angle CBA$$
$$\angle ECR = \angle EBC + \angle CBA = \angle EBA.$$

We have thus shown that triangles *CER* and *BEA* are congruent; therefore, *AE = ER*. Summing up, we have *AE = AR = ER*; consequently, Δ*AER* is equilateral.

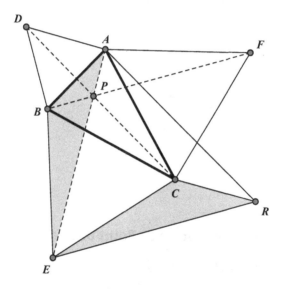

Figure 169-P

Curiosity 170. A Novel Way of Finding the Center of the Circumscribed Circle

If we look closely at the construction in Figure 170-P, we find that the lines *KD*, *LE*, and *MF* are, in fact, the perpendicular bisectors of the three sides of △*ABC*. For instance, the line *DK* joining the vertex *D* and the midpoint *K* is the bisector of the side *AB* of the equilateral △*ADB*. Similarly, *EL* is the bisector of *BC* in △*ECB*, and *FM* is the bisector of *CA* in △*FAC*.

The three lines *KD*, *LE*, and *MF* are thus concurrent at the center, *O*, of the circumscribed circle of triangle *ABC*.

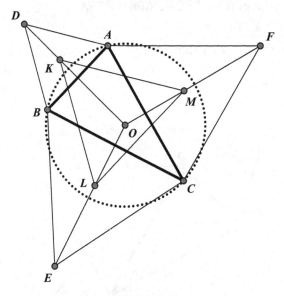

Figure 170-P

Curiosity 171. A Concurrency Point of Circles

To prove this, we need only recall the proof for Curiosity 162. In that proof, as in Figure 171-P, we defined *P* as the intersection of *AE* and *CD* and showed that a direct consequence of this definition is given by ∠*APB* = ∠*BPC* = 120°. Since ∠*BDA* = ∠*CEA* = 60°, this means that *P* is an intersection point of the circumcircle of triangles *ADB* and *BEC*. Since ∠*CPA* = 360° − ∠*APB* − ∠*BPC* = 360° − 120° − 120° = 120° and ∠*AFC* = 60° = 180° − 120° = 180° − ∠*CPA*, we then see that *P* also lies on the circumcircle of triangle *CFA*. Thus, *P* lies on all three circumcircles, which proves the circle concurrency.

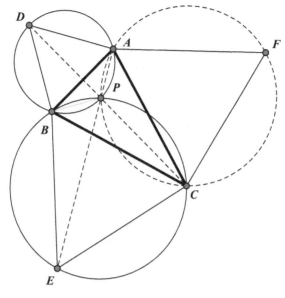

Figure 171-P

Curiosity 172. The Famous Miquel Theorem

Consider points D, E, and F on the sides of triangle ABC, as shown in Figure 172-P. Let point P be the intersection of the circumcircles of

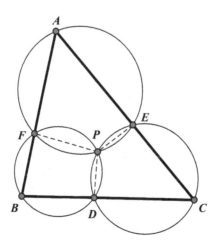

Figure 172-P

triangles *BDF* and *CED*. We need to show that this point *P* must also lie on the circumcircle of triangle *AFE*. To show this, we recall that, since the opposite angles of a cyclic quadrilateral are supplementary, $\angle FPD = 180° - \angle B$ and $\angle DPE = 180° - \angle C$. This means that we also have the following:

$$\angle EPF = 360° - \angle FPD - \angle DPE = 360° - (180° - \angle B) - (180° - \angle C)$$
$$= \angle B + \angle C = 180° - \angle A.$$

This implies that *P* lies on the circumcircle of triangle *AFE*, thus completing the proof.

Curiosity 173. Miquel's Similar Triangles

Since the circles with centers at *Q* and *R* intersect at points *P* and *F*, segment *RQ* is the perpendicular bisector of *PF*, as shown in Figure 173-P. Similarly, *SQ* is the perpendicular bisector of *PE*. Since triangles *QFP* and *QPE* are both isosceles with *QF* = *QP* = *QE*, *RQ* is also the bisector of angle $\angle FQP$, and analogously, *SQ* is the bisector of $\angle PQE$. Thus, we have $\angle Q = \angle RQP + \angle PQS = \frac{1}{2} \cdot \angle FQP + \frac{1}{2} \cdot \angle PQE = \frac{1}{2} \angle FQE$. Since *Q* is the center of the circle containing *A*, *F*, *P*, and *E*, we then have

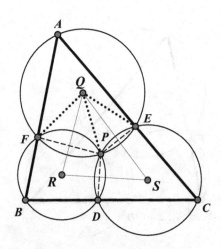

Figure 173-P

$\angle A = \frac{1}{2} \angle FQE$, as that is the relationship between a central angle and inscribed angle measured by the same arc, namely, $\overarc{FPE}$. This gives us $\angle A = \angle RQS$. Analogously, we can show that $\angle B = \angle SRQ$ and $\angle C = \angle QSR$. Thus, triangles $\triangle ABC$ and $\triangle QRS$ are similar.

Curiosity 174. The Astounding Morley's Theorem

The astonishing result of this relationship became an international challenge for mathematicians to prove. Consequently, there are now many proofs available, one of which we offer here.

We assume a randomly drawn triangle ABC as given and define $\frac{1}{3}\angle A = \alpha$, $\frac{1}{3}\angle B = \beta$, and $\frac{1}{3}\angle C = \gamma$. We then have $\alpha + \beta + \gamma = \frac{1}{3} \cdot (\angle A + \angle B + \angle C) = 60°$.

As illustrated in Figure 174-P, we now draw an equilateral triangle $E'F'G'$ and points A', B', and C', such that $\angle E'F'A' = \beta + 60°$, $\angle A'E'F' = \gamma + 60°$, $\angle F'G'B' = \gamma + 60°$, $\angle B'F'G' = \alpha + 60°$, $\angle G'E'C' = \alpha + 60°$, and $\angle C'G'E' = \beta + 60°$. We shall prove that this triangle $E'F'G'$ is identical to triangle EFG.

It then follows that:

$$\angle F'A'E' = 180° - (\angle E'F'A' + \angle A'E'F')$$
$$= 180° - (\beta + 60° + \gamma + 60°) = 60° - (\beta + \gamma) = \alpha,$$

and analogously, $\angle G'B'F' = \beta$ and $\angle E'C'G' = \gamma$.

We will now show that $\angle F'A'B' = \alpha$ and $\angle F'B'A' = \beta$.

To do this, we consider the perpendiculars $F'X$ and $F'Y$ to $A'E'$ and $B'G'$, respectively. Since $\angle A'E'F' = \angle F'G'B' = \gamma + 60°$, and $E'F' = G'F'$, right triangles $XF'E'$ and $YG'F'$ are congruent, and we then have $F'X = F'Y = d$.

We now consider the perpendicular distance h from F' to $A'B'$, measuring this distance on line segment $F'Z$. If $h < d$, then the following is true: $\angle B'A'F' < \alpha$ and $\angle F'B'A' < \beta$, and, thus, $\angle A'F'B' = 180° - \angle B'A'F' - \angle F'B'A' > 180° - (\alpha + \beta)$.

On the other hand, if $h > d$, it follows that $\angle B'A'F' > \alpha$ and $\angle F'B'A' > \beta$, and, thus, $\angle A'F'B' = 180° - \angle B'A'F' - \angle F'B'A' < 180° - (\alpha + \beta)$.

We can calculate $\angle A'F'B' = 360° - \angle E'F'A' - \angle G'F'E' - \angle B'F'G' =$ $360° - (\beta + 60°) - 60° - (\alpha + 60°) = 180° - (\alpha + \beta)$, however, and it therefore follows that $h = d$, $\angle F'A'B' = \alpha$, and $\angle F'B'A' = \beta$.

In an analogous way, we can also show that $\angle C'B'G' = \beta$, $\angle G'C'B' = \angle A'C'E' = \gamma$, and $\angle E'A'C' = \alpha$. We see that $A'E'$ and $A'F'$ trisect $\angle B'A'C'$, $B'F'$ and $B'G'$ trisect $\angle C'B'A'$, and $C'G'$ and $C'E'$ trisect $\angle A'C'B'$. Since

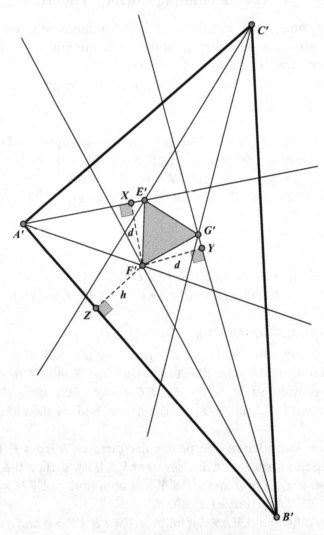

Figure 174-P

$\angle B'A'C' = 3\alpha$, $\angle C'B'A' = 3\beta$, and $\angle A'C'B' = 3\gamma$, triangle $A'B'C'$ is similar to the originally given triangle ABC. This similarity extends to all constructions, and the trisections of the interior angles of ABC therefore yield an internal triangle EFG similar to $E'F'G'$, which must also be equilateral.

Curiosity 175. Morley's Theorem Extended

The proof of this property is a nice application of the trigonometric form of Ceva's theorem (see Toolbox). We wish to show that lines AD, BE, and CF in Figure 175-P have a point in common, and by Ceva's theorem, this is equivalent to $\dfrac{\sin \angle DAC}{\sin \angle BAD} \cdot \dfrac{\sin \angle EBA}{\sin \angle CBE} \cdot \dfrac{\sin \angle FCB}{\sin \angle ACF} = 1$. By this same theorem, we know that $\dfrac{\sin \angle DAC}{\sin \angle BAD} \cdot \dfrac{\sin \angle DBA}{\sin \angle CBD} \cdot \dfrac{\sin \angle DCB}{\sin \angle ACD} = 1$, which is equivalent to $\dfrac{\sin \angle DAC}{\sin \angle BAD} = \dfrac{\sin \angle CBD}{\sin \angle DBA} \cdot \dfrac{\sin \angle ACD}{\sin \angle DCB}$ or $\dfrac{\sin \angle DAC}{\sin \angle BAD} =$

$$\dfrac{\sin \frac{1}{3} \angle CBA}{\sin \frac{2}{3} \angle CBA} \cdot \dfrac{\sin \frac{2}{3} \angle ACB}{\sin \frac{1}{3} \angle ACB}.$$

In an analogous way, we also obtain

$$\dfrac{\sin \angle DBA}{\sin \angle CBD} = \dfrac{\sin \frac{1}{3} \angle ACB}{\sin \frac{2}{3} \angle ACB} \cdot \dfrac{\sin \frac{2}{3} \angle BAC}{\sin \frac{1}{3} \angle BAC} \quad \text{and} \quad \dfrac{\sin \angle DCB}{\sin \angle ACD} = \dfrac{\sin \frac{1}{3} \angle ACB}{\sin \frac{2}{3} \angle ACB} \cdot \dfrac{\sin \frac{2}{3} \angle BAC}{\sin \frac{1}{3} \angle BAC}.$$

Substituting then yields

$$\dfrac{\sin \angle DAC}{\sin \angle BAD} \cdot \dfrac{\sin \angle DBA}{\sin \angle CBD} \cdot \dfrac{\sin \angle DCB}{\sin \angle ACD}$$
$$= \dfrac{\sin \frac{1}{3} \angle CBA}{\sin \frac{2}{3} \angle CBA} \cdot \dfrac{\sin \frac{2}{3} \angle ACB}{\sin \frac{1}{3} \angle ACB} \cdot \dfrac{\sin \frac{1}{3} \angle ACB}{\sin \frac{2}{3} \angle ACB} \cdot \dfrac{\sin \frac{2}{3} \angle BAC}{\sin \frac{1}{3} \angle BAC} \cdot \dfrac{\sin \frac{1}{3} \angle ACB}{\sin \frac{2}{3} \angle ACB} \cdot \dfrac{\sin \frac{2}{3} \angle BAC}{\sin \frac{1}{3} \angle BAC} = 1,$$

and *AD*, *BE*, and *CF* are thus concurrent.

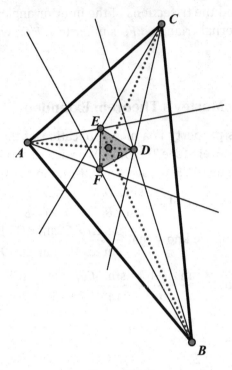

Figure 175-P

Toolbox

Introduction: The Geometry Toolbox

In the interest of providing the readership all the necessary "equipment" that might be needed for some of the proofs presented in this book, and cognizant of the fact that some readers may have forgotten some of the basics presented in the secondary geometry curriculum, we are providing this "toolbox" with some of the basic concepts that help to better appreciate the amazing relationships which we hope you will have enjoyed experiencing in the first section of this book.

First, we will review some of the facts and concepts you should be familiar with from your school experience. These results are presented without proof, as a reader's recollection should make them once again familiar. Naturally, any high school geometry book will provide further understanding of these basic concepts.

Next, we present a number of slightly more sophisticated results from basic trigonometry and somewhat more advanced Euclidean geometry that may be familiar to some readers and less so to others, but easily understandable nevertheless. These are presented with proofs, assuming only knowledge of the basic concepts from the high school geometry course. For the sake of brevity, the proofs are given for the most general cases, while special configurations may require some additional steps. Motivated readers may wish to enhance some of these aspects.

The properties in the second part of the toolbox can be considered as Wonderful Triangle Properties in their own right and could easily have been included in the core section of this book. The reason they are considered "tools" is grounded in the fact that they are generally very useful in more complex geometric proofs. Of course, this further enhances the particularly enchanting aspects of the beauty of Euclidean geometry!

A. Tools You Are Probably Familiar with from the High School Geometry Course

A1: Congruence of Triangles

Triangles can be proved congruent by showing corresponding parts equal:

- Two sides and their included angle (SAS)
- Two angles and their included side (ASA)
- Two angles and a side not included (AAS)
- Three sides (SSS)
- For right triangles: the hypotenuse and a leg (HL)

A2: Similarity of Triangles

Triangles can be proved similar by showing any of the following equivalent properties:

- The corresponding angles are equal. (AAA)
- The corresponding sides are proportional.
- Two pairs of corresponding sides are proportional and the included angles are equal.

A3: Right Triangle Properties (See Figure A3)

For the following properties, we refer to a right triangle ABC with its right angle at C. The lengths of the sides opposite vertices A, B, and C are named a, b, and c, respectively.

- The Pythagorean theorem: the sum of the squares of the legs equals the square of the hypotenuse: $a^2 + b^2 = c^2$.
- The square of the altitude h on the hypotenuse is equal to the product of the lengths of the segments p and q on the hypotenuse: $h^2 = p \cdot q$.
- The square of the length a of a leg is equal to the product of the length c of the hypotenuse and the nearest segment along the hypotenuse q, that is, $a^2 = c \cdot q$, also $b^2 = c \cdot p$.

- The median *CM* to the hypotenuse of a right triangle is half the length *c* of the hypotenuse.

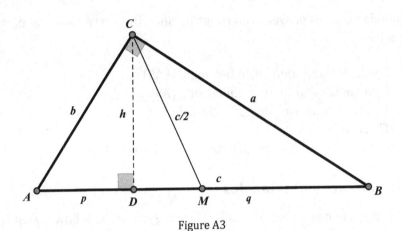

Figure A3

A4: Angles Related to a Circle (see Figure A4)

(Note that these relationships are given in two different common notations, among the many in use internationally, in the hope that one of these will be sufficiently familiar for any reader.)

- An inscribed angle: $\angle A = \frac{1}{2}\overset{\frown}{BC}$ (or $\angle CAB = \frac{1}{2}\angle COB$)
- An angle formed by 2 secants: $\angle F = \frac{1}{2}\left(\overset{\frown}{BC} - \overset{\frown}{DE}\right)$ (or $\angle CFB = \frac{1}{2}(\angle COB - \angle EOD)$)
- An angle formed by 2 tangents: $\angle J = \frac{1}{2}\left(\overset{\frown}{HAK} - \overset{\frown}{HCK}\right) = 180° - \overset{\frown}{HCK}$ (or $\angle HJK = \frac{1}{2}(180° - \angle KOH)$)
- An angle formed by a secant and a tangent: $\angle L = \frac{1}{2}\left(\overset{\frown}{CK} - \overset{\frown}{NK}\right)$ (or $\angle KLC = \frac{1}{2}(\angle KOC - \angle NOK)$)
- An angle formed by 2 chords intersecting inside the circle: $\angle AME = \frac{1}{2}\left(\overset{\frown}{AE} + \overset{\frown}{BC}\right)$ (or $\angle AME = \frac{1}{2}(\angle AOE + \angle COB)$)
- An angle formed by a chord and a tangent: $\angle JKC = \frac{1}{2}\overset{\frown}{KC} = \angle KAC$ (or $\angle JKC = \frac{1}{2}\angle KOC = \angle KAC$)

- Opposite angles in a circle (i.e., opposite angles of an inscribed quadrilateral): $\angle BCK = 180° - \angle KAB$

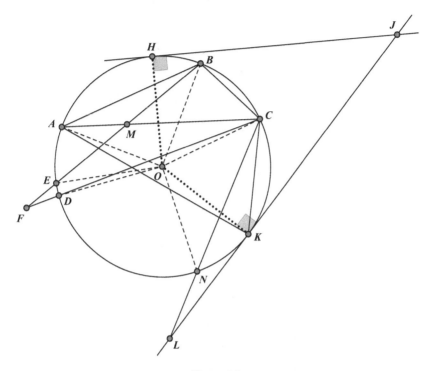

Figure A4

A5: Tangents, Secants, and Chords: Segments of a Circle (see Figure A5)

- From the same external point, the tangent is a mean proportional between the entire secant and its external segment: $\dfrac{e+f}{h} = \dfrac{h}{f}$, or $h^2 = f(e + f)$.
- From the same external point, for two secants: the product of the length of a secant and its external segment is equal to the product of the other secant and its external segment: $(e + f) f = (g + k) k$.
- For two chords intersecting inside a circle, the product of the segments of one chord is equal to the product of the segments of the other chord: $a \cdot b = c \cdot d$.

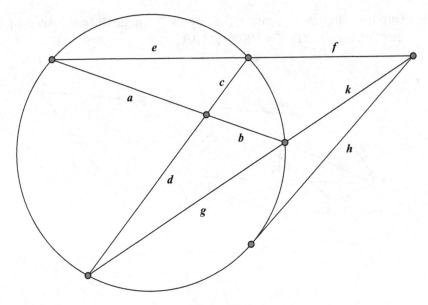

Figure A5

A6: The Law of Sines and the Law of Cosines

In Figure A6, for triangle *ABC*, we let *a*, *b*, and *c* denote the lengths of the sides opposite the vertices *A*, *B*, and *C*, respectively. Then $\frac{a}{\sin\angle A} = \frac{b}{\sin\angle B} = \frac{c}{\sin\angle C}$. This is known as the *Law of Sines*.

Also, $c^2 = a^2 + b^2 - 2ab\cos\angle ACB$, and analogously $a^2 = b^2 + c^2 - 2bc\cos\angle BAC$ and $b^2 = c^2 + a^2 - 2ca\cos\angle CBA$. This is known as the *Law of Cosines*.

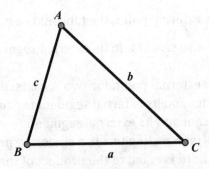

Figure A6

A7: Angle Sum and Difference Identities

It is often useful to express the sine or cosine of the sum or difference of two angles in terms of the individual sines and cosines. To this purpose, we can use the following identities:

$$\sin(\alpha + \beta) = \sin\alpha\cos\beta + \cos\alpha\sin\beta$$
$$\sin(\alpha - \beta) = \sin\alpha\cos\beta - \cos\alpha\sin\beta$$
$$\cos(\alpha + \beta) = \cos\alpha\cos\beta - \sin\alpha\sin\beta$$
$$\cos(\alpha - \beta) = \cos\alpha\cos\beta + \sin\alpha\sin\beta$$

B. Less Familiar Tools—However, Useful and Fascinating

B1. Interior Angle Bisector in a Triangle

The interior angle bisector of a triangle partitions the side to which it is drawn proportional to the two adjacent sides of the bisected angle. In Figure B1, line AD is a bisector of angle $\angle BAC$; therefore, $\dfrac{DC}{DB} = \dfrac{AC}{AB}$.

Proof: In Figure B1, line AD is the bisector of $\angle BAC$ of $\triangle ABC$. The line parallel to AD through point B intersects the extension of side CA at point E. To make matters simpler, we have marked four angles in the figure, all of which are equal. We have $\angle ABE = \angle BAD$, as these are alternate-interior angles of the parallel lines AD and BE. Furthermore, $\angle BAE = \angle DAC$ because they are corresponding angles of these parallel lines, and since AD is the bisector of $\angle BAC$, we also have $\angle BAD = \angle DAC$. Therefore, by substitution, we get $\angle BAE = \angle ABE$. This yields isosceles triangle ABE, where $AE = AB$. Since triangles EBC and ADC are similar, we have $\dfrac{DC}{DB} = \dfrac{AC}{AE}$. Now, by substituting AB for AE, we get the desired result, namely, $\dfrac{DC}{DB} = \dfrac{AC}{AB}$.

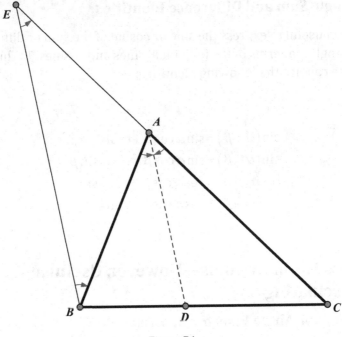

Figure B1

B2. Exterior Angle Bisector in a Triangle

The exterior angle bisector of a triangle partitions the line containing the opposite side to which it is drawn into segments proportional to the two adjacent sides of the bisected angle. In Figure B2, line AD is a bisector of angle $\angle EAB$; therefore, $\dfrac{DC}{DB} = \dfrac{AC}{AB}$.

Proof: In figure B2, line AD is the bisector of exterior angle $\angle EAB$ of $\triangle ABC$. The line parallel to AD through point B intersects side CA at point E. As in the previous proof, we have again marked four angles in the figure, all of which are equal.

We have $\angle ABE = \angle DAB$, as these are alternate-interior angles of the parallel lines AD and BE. Furthermore, $\angle EAD = \angle AEB$ because they are corresponding angles of these parallel lines, and since AD is the bisector of $\angle EAB$, we also have $\angle EAD = \angle DAB$. Therefore, by substitution we get $\angle AEB = \angle EBA$. This yields isosceles triangle ABE, where $AE =$

AB. Since triangles *EBC* and *ADC* are similar, we have $\dfrac{DC}{DB} = \dfrac{AC}{AE}$. By substituting *AB* for *AE*, we once again get the desired result, namely, $\dfrac{DC}{DB} = \dfrac{AC}{AB}$.

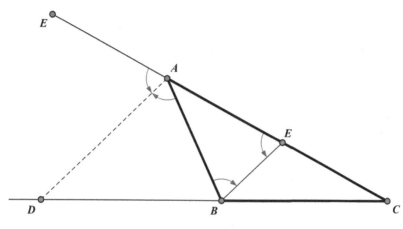

Figure B2

B3. Menelaus' Theorem

In Figure B3a, we have triangle *ABC* with three points *D*, *E*, and *F* on sides *BC* (extended), *CA*, and *AB*, respectively. Menelaus' theorem

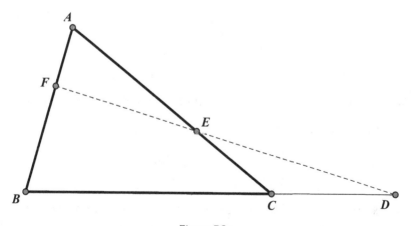

Figure B3a

states that the three points *D*, *E*, and *F* are collinear (lie on the same line), if, and only if, the alternate segments they determine on the sides have equal products. Expressed as an equation, this can be written as *AF·BD·CE = AE·BF·CD*.

Proof: We will divide the proof of this into several steps.

First of all, we assume that *D*, *E*, and *F* are collinear, and show that *AF·BD·CE = AE·BF·CD* must then follow. In Figure B3b, we draw a line containing point *C*, parallel to *AB*, and intersecting *DEF* at point *P*.

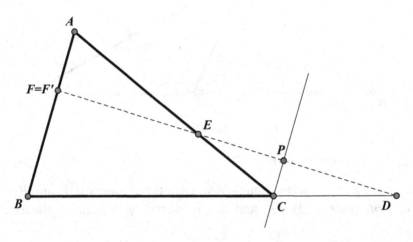

Figure B3b

Since triangles *PCD* and *FBD* are similar, we have $\frac{CP}{BF} = \frac{CD}{BD}$, or $CP = \frac{BF \cdot CD}{BD}$. Furthermore, triangles *PEC* and *FEA* are also similar, and we, therefore, have $\frac{CP}{AF} = \frac{CE}{AE}$, or $CP = \frac{AF \cdot CE}{AE}$. This gives us $\frac{BF \cdot CD}{BD} = \frac{AF \cdot CE}{AE}$, which is equivalent to *AF·BD·CE = AE·BF·CD*. This proves the first part of Menelaus' theorem.

Next, we assume that *AF·BD·CE = AE·BF·CD*, and show that points *D*, *E*, and *F* must be collinear. We let the intersection point of *AB* and

DE be the point F'. Then we have to prove $F' = F$. Since points *D*, *E*, and F' are collinear, we have just shown that $AF' \cdot BD \cdot CE = AE \cdot BF' \cdot CD$ holds, which we can also write as $\dfrac{AE}{CE} \cdot \dfrac{BF'}{AF'} \cdot \dfrac{CD}{BD} = 1$. Since we are assuming that $AF \cdot BD \cdot CE = AE \cdot BF \cdot CD$, which we can also write as $\dfrac{AE}{CE} \cdot \dfrac{BF}{AF} \cdot \dfrac{CD}{BD} = 1$, we, therefore, have $\dfrac{BF'}{AF'} = \dfrac{BF}{AF}$, and so we have shown that $F = F'$.

As a last step, we now consider the configuration illustrated in Figure B3c.

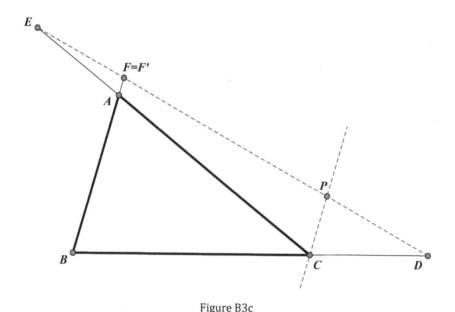

Figure B3c

Here, we have triangle *ABC* with three points *D*, *E*, and *F* on the extended sides *BC*, *CA*, and *AB*, respectively. In this situation, we can also show that the three points *D*, *E*, and *F* are collinear if, and only if, $AF \cdot BD \cdot CE = AE \cdot BF \cdot CD$. The proof in this case is identical to that previously shown.

We see that Menelaus' theorem can be stated in a more general fashion as follows: We are given a triangle *ABC*. Points *D*, *E*, and *F* lie on the lines *BC*, *CA*, and *AB*, respectively. These points are collinear, if, and only if, $AF \cdot BD \cdot CE = AE \cdot BF \cdot CD$, or $\dfrac{AE}{CE} \cdot \dfrac{BF}{AF} \cdot \dfrac{CD}{BD} = 1$.

B4. Ceva's Theorem

A common topic in triangle geometry concerns lines joining the vertices of a triangle with points on their opposite sides. Such lines are often called *cevians*, named after the Italian mathematician Giovanni Ceva (1647–1734), who published his famous theorem in 1678. The result is very powerful and leads to many geometric wonders. Ceva's theorem states that three cevians are concurrent, if, and only if, the product of the lengths of the alternate segments made by the points of contact on the sides are equal. We see this illustrated in Figure B4a, where triangle *ABC* has cevians *AD*, *BE*, and *CF* meeting the sides *BC*, *CA*, and *AB* in points *D*, *E*, and *F*, respectively. In the figure, they are shown to meet at a common point *P*. According to Ceva's theorem, the cevians *BC*, *CA*, and *AB* intersect in a common point *P* if, and only if, $AF \cdot BD \cdot CE = AE \cdot BF \cdot CD$, or, equivalently, $\dfrac{AE}{CE} \cdot \dfrac{BF}{AF} \cdot \dfrac{CD}{BD} = 1$.

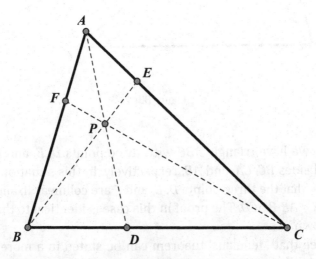

Figure B4a

Proof: As we did for the proof of Menelaus's theorem, we will divide this proof into steps. First, we assume that the three cevians BC, CA, and AB meet in a common point P. We will show that $\dfrac{AE}{CE}\cdot\dfrac{BF}{AF}\cdot\dfrac{CD}{BD}=1$.

In order to show this, we consider Figure B4b. Here, starting from the configuration in Figure B4a, we have added a line parallel to BC through A. This line intersects BE extended at point R and CF extended at point S.

The parallel lines enable us to establish several pairs of similar triangles, and relationships between the lengths of their sides result from these similarities. We have

$$\triangle AER \sim \triangle CEB,\text{ and therefore, }\frac{AE}{CE}=\frac{AR}{BC}, \qquad\text{(I)}$$

$$\triangle BCF \sim \triangle ASF,\text{ and therefore, }\frac{BF}{AF}=\frac{BC}{AS}, \qquad\text{(II)}$$

$$\triangle CPD \sim \triangle SPA,\text{ and therefore, }\frac{CD}{AS}=\frac{DP}{AP}, \qquad\text{(III)}$$

$$\triangle BDP \sim \triangle RAP,\text{ and therefore, }\frac{BD}{AR}=\frac{DP}{AP}, \qquad\text{(IV)}$$

From (III) and (IV) we get $\dfrac{CD}{AS}=\dfrac{BD}{AR}$, or $\dfrac{CD}{BD}=\dfrac{AS}{AR}$. $\qquad$ (V)

Now, by multiplying (I), (II), and (V), we obtain our desired result:

$$\frac{AE}{CE}\cdot\frac{BF}{AF}\cdot\frac{CD}{BD}=\frac{AR}{BC}\cdot\frac{BC}{AS}\cdot\frac{AS}{AR}=1.$$

Next, we assume that points D, E, and F are given on the sides BC, CA, and AB, respectively, of triangle ABC, with $\dfrac{AE}{CE}\cdot\dfrac{BF}{AF}\cdot\dfrac{CD}{BD}=1$. We wish to

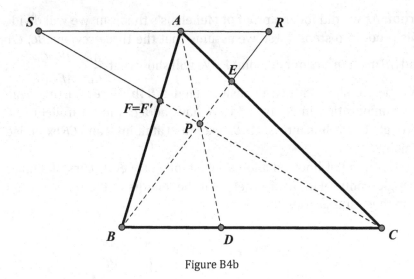

Figure B4b

show that line segments AD, BE, and CF then have a common point P. Suppose AD and BE intersect at P. Draw CP and call its intersection with AB point F'. Since AD, BE, and CF' are concurrent, we can use the part of Ceva's theorem we have already proved to state the following: $\dfrac{AE}{CE} \cdot \dfrac{BF'}{AF'} \cdot \dfrac{CD}{BD} = 1$. Since our hypothesis stated that $\dfrac{AE}{CE} \cdot \dfrac{BF}{AF} \cdot \dfrac{CD}{BD} = 1$, we obtain $\dfrac{BF'}{AF'} = \dfrac{BF}{AF}$, and thus $F = F'$, proving the concurrency.

B5. The Trigonometric Version of Ceva's Theorem

There is an interesting variation to Ceva's theorem, which was discovered by the French mathematician Lazare Carnot (1753–1823). Referring, once again, to Figure B4a, this version states that cevians AD, BE, and CF meet in a common point P if, and only if, $\dfrac{\sin \angle BAD}{\sin \angle DAC} \cdot \dfrac{\sin \angle CBE}{\sin \angle EBA} \cdot \dfrac{\sin \angle ACF}{\sin \angle FCB} = 1$.

Proof: This statement is equivalent to the usual version of the theorem, and we will show this by applying the law of sines several times as follows:

Referring to Figure B4a, we consider triangle ADC. In this triangle, the law of sines gives us $\dfrac{CD}{CA} = \dfrac{\sin \angle DAC}{\sin \angle CDA}$ or $CD = CA \cdot \dfrac{\sin \angle DAC}{\sin \angle CDA}$.
Analogous considerations in triangles BCE, BEA, CAF, CFB, and ABD give us $CE = BC \cdot \dfrac{\sin \angle CBE}{\sin \angle BEC}$, $AE = AB \cdot \dfrac{\sin \angle EBA}{\sin \angle AEB}$, $AF = CA \cdot \dfrac{\sin \angle ACF}{\sin \angle CFA}$,
$BF = BC \cdot \dfrac{\sin \angle FCB}{\sin \angle BFC}$, and $BD = AB \cdot \dfrac{\sin \angle BAD}{\sin \angle ADB}$, respectively.

We have already established that AD, BE, and CF meet in a common point P if, and only if, $\dfrac{AE}{CE} \cdot \dfrac{BF}{AF} \cdot \dfrac{CD}{BD} = 1$. Substitution gives us

$$\frac{AE}{CE} \cdot \frac{BF}{AF} \cdot \frac{CD}{BD}$$
$$= \left(\frac{AB \cdot \sin \angle EBA \cdot \sin \angle BEC}{BC \cdot \sin \angle CBE \cdot \sin \angle AEB} \right) \cdot \left(\frac{BC \cdot \sin \angle FCB \cdot \sin \angle CFA}{CA \cdot \sin \angle ACF \cdot \sin \angle BFC} \right) \left(\frac{CA \cdot \sin \angle DAC \cdot \sin \angle ADB}{AB \cdot \sin \angle BAD \cdot \sin \angle CDA} \right)$$
$$= \left(\frac{\sin \angle EBA \cdot \sin \angle BEC}{\sin \angle CBE \cdot \sin \angle AEB} \right) \cdot \left(\frac{\sin \angle FCB \cdot \sin \angle CFA}{\sin \angle ACF \cdot \sin \angle BFC} \right) \cdot \left(\frac{\sin \angle DAC \cdot \sin \angle ADB}{\sin \angle BAD \cdot \sin \angle CDA} \right).$$

Since we have $\angle ADB = 180° - \angle CDA$, $\angle BEC = 180° - \angle AEB$ and $\angle CFA = 180° - \angle BFC$, we also have $\sin \angle ADB = \sin \angle CDA$, $\sin \angle BEC = \sin \angle AEB$, and $\sin \angle CFA = \sin \angle BFC$, and cancelling the common factors gives us:

$$\frac{AE}{CE} \cdot \frac{BF}{AF} \cdot \frac{CD}{BD} = \frac{\sin \angle EBA}{\sin \angle CBE} \cdot \frac{\sin \angle FCB}{\sin \angle ACF} \cdot \frac{\sin \angle DAC}{\sin \angle BAD}.$$

We see that $\dfrac{AE}{CE} \cdot \dfrac{BF}{AF} \cdot \dfrac{CD}{BD} = 1$ is equivalent to $\dfrac{\sin \angle EBA}{\sin \angle CBE} \cdot \dfrac{\sin \angle FCB}{\sin \angle ACF} \cdot \dfrac{\sin \angle DAC}{\sin \angle BAD} = 1$, which we can also write as $\dfrac{\sin \angle BAD}{\sin \angle DAC} \cdot \dfrac{\sin \angle CBE}{\sin \angle EBA} \cdot \dfrac{\sin \angle ACF}{\sin \angle FCB} = 1$, and the validity of this equation is, thus, equivalent with the concurrence of AD, BE, and CF.

B6. Ceva's Theorem Extended

Ceva's theorem also holds true when the cevians intersect outside the triangle. In Figure B6a, the cevians *AD*, *BE*, and *CF* meet the sides *BC*, *CA*, and *AB* in points *D*, *E*, and *F*, respectively. Here, the configuration is such that they intersect outside the triangle. In this case, it is also true that *AD*, *BE*, and *CF* meet in a common point *P* if, and only if,

$$\frac{AE}{CE} \cdot \frac{BF}{AF} \cdot \frac{CD}{BD} = 1.$$

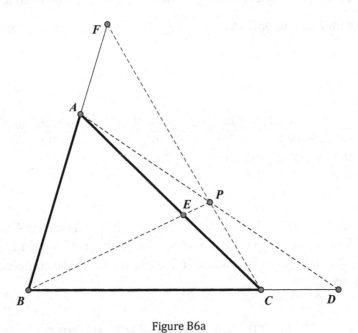

Figure B6a

Proof: The proof in this case is identical to the proof when *P* is in the interior of *ABC*. In Figure B6b, we have added points *R* and *S* in the same way we did in Figure B4b, and the proof from B4 is identical for this configuration.

B7. Desargues's Theorem

A very useful geometric relationship, discovered by French mathematician Gérard Desargues (1591–1661), was first presented in a book

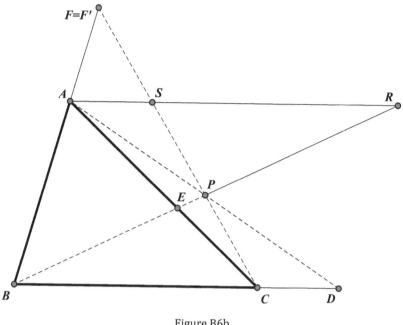

Figure B6b

entitled *Manière universelle de M. Desargues, pour pratiquer la perspective.* It involves two triangles placed so that the three lines joining corresponding vertices are concurrent. Remarkably, when this is the case, the pairs of corresponding sides intersect in three collinear points. In figure B7, ΔABC and $\Delta A'B'C'$ are situated in such a way that the lines joining the corresponding vertices, AA', BB', and CC', are concurrent. The pairs of corresponding sides, therefore, intersect in three collinear points. In other words, the point Q, at which BC and $B'C'$ intersect, the point R, at which CA and $C'A'$ intersect, and the point S, at which AB and $A'B'$ intersect, are collinear.

The converse is also true; if the two triangles ABC and $A'B'C'$ are situated in such a way that points Q, R, and S lie on a common line, lines AA', BB', and CC' contain a common point P.

Proof: We shall prove Desargues's theorem by applying Menelaus' theorem several times. Consider line QBC as a transversal of $\Delta PB'C'$. By Menelaus' theorem, we have

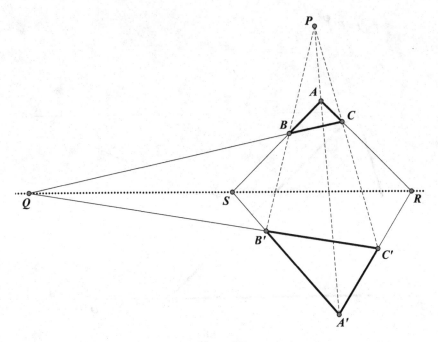

Figure B7

$$\frac{PB}{BB'} \cdot \frac{B'Q}{QC'} \cdot \frac{C'C}{CP} = 1. \tag{I}$$

Similarly, considering SBA as a transversal of $\Delta PB'A'$, we have

$$\frac{PA}{AA'} \cdot \frac{A'S}{SB'} \cdot \frac{B'B}{BP} = 1, \tag{II}$$

and considering RCA as a transversal of $\Delta PA'C'$, we have

$$\frac{PC}{CC'} \cdot \frac{C'R}{RA'} \cdot \frac{A'A}{AP} = 1 \tag{III}$$

By multiplying (I), (II), and (III), we get:

$$\left(\frac{PB}{BB'} \cdot \frac{B'Q}{QC'} \cdot \frac{C'C}{CP}\right)\left(\frac{PA}{AA'} \cdot \frac{A'S}{SB'} \cdot \frac{B'B}{BP}\right)\left(\frac{PC}{CC'} \cdot \frac{C'R}{RA'} \cdot \frac{A'A}{AP}\right) = 1,$$

or

$$\frac{B'Q}{QC'} \cdot \frac{A'S}{SB'} \cdot \frac{C'R}{RA'} = 1.$$

Thus, by Menelaus' theorem applied to $\triangle A'B'C'$, we have points A', B', and C' collinear.

It now remains to prove that the converse is true. We assume that ABC and $A'B'C'$ are situated such that points Q, R, and S lie on a common line, and wish to show that lines AA', BB', and CC' contain a common point P. Somewhat surprisingly, this is an immediate consequence of what we have just proved.

Consider triangles $SB'B$ and RCC'. Because of our assumptions, we know that lines BC, $B'C'$, and RS have a common point, namely, Q. From the version of Desargues's theorem that we have just proved, we know that the pairs of corresponding sides of $SB'B$ and RCC' intersect in three collinear points. In other words, the point P, at which BB' and CC' intersect, the point A, at which BS and CR intersect, and the point A', at which $B'S$ and $C'R$ intersect, are collinear. Lines AA', BB', and CC', therefore, all pass through the common point P.

B8. Stewart's Theorem

A helpful formula for determining the length of a cevian in a given triangle is generally credited to the Scottish mathematician Matthew Stewart (1717–1785), although it may have been originally discovered by his teacher, the famous Scottish geometer Robert Simson (1687–1768).

In Figure B8a, we have triangle ABC with $BC = a$, $CA = b$, and $AB = c$. Point D divides BC into two segments, with $BD = m$, $CD = n$, and $AD = d$. Stewart's theorem then states

$$c^2 n + b^2 m = a(d^2 + mn).$$

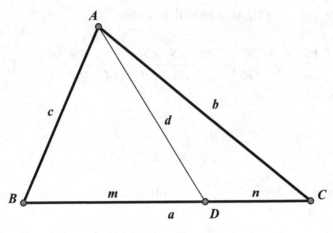

Figure B8a

Proof: In Figure B8b, we have points *A*, *B*, *C*, and *D* as in Figure B8a. Furthermore, we have drawn the altitude *AE = h* and let *DE = p*.

We first apply the Pythagorean theorem to triangle *ABE* to obtain $AB^2 = AE^2 + BE^2$.

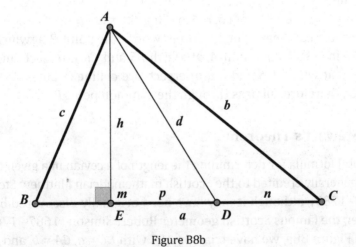

Figure B8b

Since *BE = m − p*, this can be written as

$$c^2 = h^2 + (m - p)^2. \tag{I}$$

Applying the Pythagorean theorem to triangle AED gives us $AE^2 = AD^2 - DE^2$, or $h^2 = d^2 - p^2$. Substituting for h^2 in equation (I), we then obtain $c^2 = d^2 - p^2 + (m - p)^2 = d^2 - p^2 + m^2 - 2mp + p^2$, and thus,

$$c^2 = d^2 + m^2 - 2mp. \tag{II}$$

Similarly, in triangle ACE, we have $AC^2 = AE^2 + CE^2$, and since $CE = n + p$,

$$b^2 = h^2 + (n + p)^2. \tag{III}$$

Substituting for h^2 in (III) then gives us $b^2 = d^2 - p^2 + (n + p)^2$, and thus,

$$b^2 = d^2 + n^2 + 2np. \tag{IV}$$

Now, multiply equation (II) by n to get

$$c^2n = d^2n + m^2n - 2mnp, \tag{V}$$

and multiply equation (IV) by m to get

$$b^2m = d^2m + mn^2 + 2mnp. \tag{VI}$$

Adding (V) and (VI), we then have $c^2m + b^2n = d^2m + d^2n + m^2n + mn^2 + 2mnp - 2mnp$. Therefore, $c^2m + b^2n = d^2(m + n) + mn(m + n)$.

Since $m + n = a$, we have $c^2m + b^2n = d^2a + mna = a(d^2 + mn)$, which is the relationship we set out to justify.

B9. Theorem of Apollonius

The theorem of Apollonius states that the sum of the squares of any two sides of a triangle is equal to twice the sum of the squares of the median to the third side and half the third side. In triangle ABC, where AD is the median to side BC, this can be stated as

$$AB^2 + AC^2 = 2(AD^2 + BD^2).$$

Proof: In Figure B9, we have drawn point E on BC, such that AE is the altitude of ABC. To simplify matters, we will let $BD = CD = x$, $DE = y$, and $AE = h$. This then gives us

$$BE = (x - y) \text{ and } CE = (x + y).$$

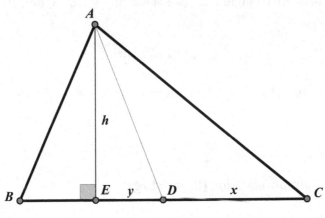

Figure B9

We now apply the Pythagorean theorem to right triangles ABE and ACE to get $AB^2 = AE^2 + BE^2$, and $AC^2 = AE^2 + CE^2$. Adding these equations, we then get $AB^2 + AC^2 = AE^2 + BE^2 + AE^2 + CE^2 = h^2 + (x - y)^2 + h^2 + (x + y)^2$, or $AB^2 + AC^2 = h^2 + x^2 - 2xy + y^2 + h^2 + x^2 + 2xy + y^2 = 2(h^2 + x^2 + x^2)$.

Since $AD^2 = h^2 + y^2$, and $BD^2 = x^2$, this yields $AB^2 + AC^2 = 2(AD^2 + BD^2)$.

B10. Ptolemy's Theorem

The product of the lengths of the diagonals of a cyclic quadrilateral equals the sum of the products of the lengths of the pairs of opposite sides. In other words, in Figure B10, we have for cyclic quadrilateral $AC \cdot BD = AB \cdot CD + BC \cdot DA$.

Proof: In Figure B10, quadrilateral $ABCD$ is inscribed in circle O. A line is drawn through A to meet CD extended at P, so that $\angle BAC = \angle DAP$. Since quadrilateral $ABCD$ is cyclic, $\angle CBA$ is supplementary to $\angle ADC$.

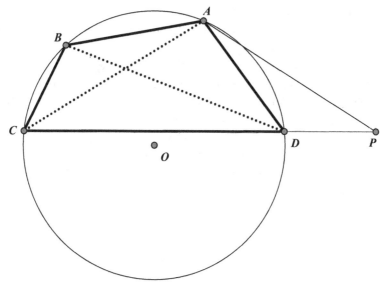

Figure B10

Since $\angle PDA$ is also supplementary to $\angle ADC$, we have $\angle CBA = \angle PDA$, and because of the way point P was constructed, triangles BCA and DPA are similar. We then have $\dfrac{AB}{DA} = \dfrac{BC}{DP}$, or $DP = \dfrac{BC \cdot DA}{AB}$. Furthermore, we also get $\dfrac{AB}{DA} = \dfrac{AC}{AP}$. Since $\angle BAD = \angle BAC + \angle CAD = \angle DAP + \angle CAD = \angle CAP$, triangles ABD and ACP are also similar. We then also have $\dfrac{BD}{CP} = \dfrac{AB}{AC}$, or $CP = \dfrac{AC \cdot BD}{AB}$. We know that $CP = CD + DP$, and substituting for CP and DP, we get $\dfrac{AC \cdot BD}{AB} = CD + \dfrac{BC \cdot DA}{AB}$, or $AC \cdot BD = AB \cdot CD + BC \cdot DA$.

B11. Isometries: Reflection and Rotation

In the course of a geometric proof, it is often useful to define a point transformation that maps a configuration of points onto a congruent configuration. Such transformations are known as *isometries*. The defining characteristic of an isometry is the fact that it maps every point in the plane onto a unique point in such a way that the distance between any two points is equal to the distance between their images. This means that any geometric figure will always be mapped onto a congruent figure by the isometry.

Two important examples of isometries are *reflections* and *rotations*. These are illustrated in the following figures. In Figure B11a, we see a reflection defined by line *l*. We say that a point *P* is *reflected* about line *l* onto a point *P′*, where *P′* is the point *symmetric* to *P* with respect to *l*. From the point *P*, a line perpendicular to line *l* through point *P* intersects line *l* in point *L*. Point *P′* is then defined as the point on this line with $LP = LP'$ and $P \neq P'$. (Note that, if a point *P* lies on line *l*, we define $P = P'$.)

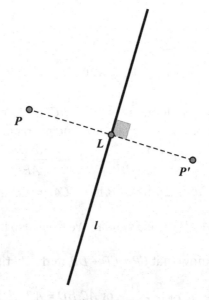

Figure B11a

In Figure B11b, we see a rotation defined by center *O* and angle *θ*. We say that a point *P* is *rotated* about *O* through angle measure *θ* onto a point *P′*. From the point *P*, line segment *OP* is drawn. Point *P′* is then defined as the point with $\angle POP' = \theta$ and $OP = OP'$. (Note that, for the sake of completeness, we define that *O* is rotated onto itself, independent of the size of angle *θ*.)

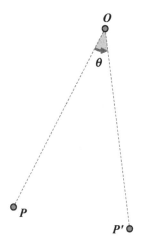

Figure B11b

In order to prove that these transformations are indeed iso-metries, we must show that all line segments are transformed onto line segments of equal length.

In Figure B11c, we see a line segment PQ reflected about line l onto a line segment $P'Q'$. In order to prove that $PQ = P'Q'$, we draw lines MP and MP', where M is the intersection of line l and QQ'. Since $LP = LP'$ and $PP' \perp LM = l$, triangles LPM and LMP' are congruent, and we have $MP = MP'$ and $\angle LMP = \angle P'ML$. This allows us to note $MQ = MQ'$, $\angle PMQ = \angle LMQ - \angle LMP = \angle Q'ML - \angle P'ML = \angle Q'MP'$, and $MP = MP'$, which shows us that triangles MPQ and $MQ'P'$ are also congruent. From this, we see that $PQ = P'Q'$ holds.

Note that we assumed here that P and Q both lie on the same side of l. If they lie on opposite sides of line l, or one or both lie on line l, the proof must be modified a bit, but this is left to the interested reader as an easy exercise.

Similarly, in Figure B11d, we see a line segment PQ rotated about O through angle measure θ onto a line segment $P'Q'$. To prove that $PQ = P'Q'$, we consider triangles OPQ and $OP'Q'$. Since $OP = OP'$, $OQ = OQ'$, and $\angle POQ = \angle POP' - \angle QOP' = \theta - \angle QOP' = \angle QOQ' - \angle QOP' = \angle P'OQ'$, we see that triangles OPQ and $OP'Q'$ are congruent. From this, we again see that $PQ = P'Q'$. As above, we note that the details of this

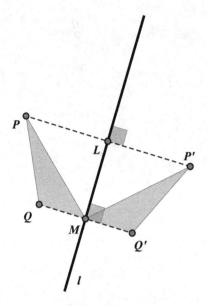

Figure B11c

depend on the relative positions of the points, but fully analogous proofs will work for any other positions, and these are left to the interested reader.

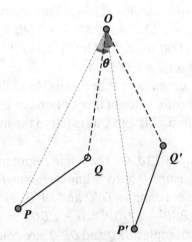

Figure B11d

B12. Homothety and Similarity

A slightly more general point transformation that is also frequently useful is one that maps a configuration of points onto a geometrically similar configuration. Such transformations are known as *similarity transformations*. The defining characteristic of a similarity transformation is the fact that it maps every point in the plane onto a unique point in such a way that the distance between any two points, multiplied by a constant, is equal to the distance between their maps. This means that any geometric figure will always be mapped onto a similar figure by such a transformation. An important example of a similarity transformation is *homothety*. This is illustrated in Figure B12a.

A *homothety* is a mapping of points P in a plane onto corresponding points P' for which there exists a fixed point O and a fixed factor a (that is, a real number $a \neq 0$) such that $SP' = a \cdot SP$ for all points P.

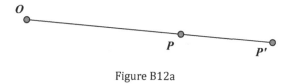

Figure B12a

The most important property of a homothety is that it maps any line segment PQ onto a line segment $P'Q'$ with $PQ \| P'Q'$ and $P'Q' = a \cdot PQ$. In order to prove this, we consider Figure B12b. Since $OP' = a \cdot OP$ and

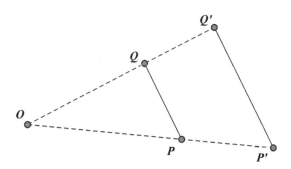

Figure B12b

$OQ' = a \cdot OQ$, and triangles OPQ and $OP'Q'$ share a common angle in O, they are similar. We therefore have $a = \dfrac{OP'}{OP} = \dfrac{OQ'}{OQ} = \dfrac{P'Q'}{PQ}$, whereupon it follows that $P'Q' = a \cdot PQ$. Furthermore, since $\angle QPO = \angle Q'P'O$, we also have $PQ \| P'Q'$.

B13. Polarity on Circles

We define polarity with respect to a circle with center O and radius r in the following way. For any point P (other than O) there exists a unique point Q on the ray OP with $OP \cdot OQ = r^2$. We refer to the line p through Q, and perpendicular to OP, as the *polar* of P, and to P as the *pole* of p. This is illustrated in Figure B13.

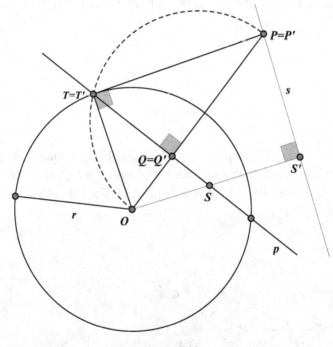

Figure B13

Polarity with respect to a circle has many interesting properties. Some of these are as follows.

If P is chosen outside the circle O, as shown in Figure B13, and we define T as one of the intersections of the circle with the polar p, then PT is a tangent of the circle O. Also, if a point S is selected on the polar p, the pole P lies on the polar s of S with respect to circle O.

Proof: By the definition of the polarity, we have $OP \cdot OQ = r^2$. We now consider the point T' on the circle O such that $PT' \perp OT'$. We then know that PT' is a tangent of circle O, and that $OP^2 = OT'^2 + PT'^2$. Next, we choose point Q' on OP such that $T'Q' \perp OP$. Since right triangles $T'OP$ and $Q'T'O$ have a common angle at O, they are similar, and we have $\dfrac{OQ'}{OT'} = \dfrac{OT'}{OP}$. This is equivalent to $OP \cdot OQ' = OT'^2 = r^2$, and we see that $Q = Q'$, which also implies $T = T'$. This means that PT is a tangent of the circle O, as we had claimed.

We now will show that the polar s of S contains the pole P of p, if S is chosen on p. The proof of this part is surprisingly similar to the proof of the first part. We assume that S' is the point on the ray OS with $OS \cdot OS' = r^2$, and P' the intersection of ray OP with the polar s of S. As before, we once again have two similar right triangles with a common angle in O, namely, QOS and $S'P'O$. This means that $\dfrac{OP'}{OS} = \dfrac{OS'}{OQ}$, which is equivalent to $OP' \cdot OQ = OS \cdot OS' = r^2$. This means that $P = P'$, and P therefore lies on s, as we had set out to show.

Index

Printed in the United States
by Baker & Taylor Publisher Services